Mathematical Theory
of Bayesian Statistics

Mathematical Theory of Bayesian Statistics

Sumio Watanabe

CRC Press
Taylor & Francis Group
Boca Raton London New York

CRC Press is an imprint of the
Taylor & Francis Group, an **informa** business

A CHAPMAN & HALL BOOK

CRC Press
Taylor & Francis Group
6000 Broken Sound Parkway NW, Suite 300
Boca Raton, FL 33487-2742

First issued in paperback 2020

ISBN-13: 978-1-4822-3806-8 (hbk)
ISBN-13: 978-0-367-73481-7 (pbk)

Visit the Taylor & Francis Web site at
http://www.taylorandfrancis.com

and the CRC Press Web site at
http://www.crcpress.com

Contents

Preface   ix

1   Definition of Bayesian Statistics   **1**
   1.1   Bayesian Statistics   2
   1.2   Probability Distribution   4
   1.3   True Distribution   7
   1.4   Model, Prior, and Posterior   9
   1.5   Examples of Posterior Distributions   11
   1.6   Estimation and Generalization   17
   1.7   Marginal Likelihood or Partition Function   21
   1.8   Conditional Independent Cases   25
   1.9   Problems   28

2   Statistical Models   **35**
   2.1   Normal Distribution   35
   2.2   Multinomial Distribution   41
   2.3   Linear Regression   48
   2.4   Neural Network   53
   2.5   Finite Normal Mixture   56
   2.6   Nonparametric Mixture   59
   2.7   Problems   63

3   Basic Formula of Bayesian Observables   **67**
   3.1   Formal Relation between True and Model   67
   3.2   Normalized Observables   77
   3.3   Cumulant Generating Functions   80
   3.4   Basic Bayesian Theory   85
   3.5   Problems   94

4 Regular Posterior Distribution   **99**
  4.1   Division of Partition Function 99
  4.2   Asymptotic Free Energy 107
  4.3   Asymptotic Losses 111
  4.4   Proof of Asymptotic Expansions 118
  4.5   Point Estimators 123
  4.6   Problems 126

5 Standard Posterior Distribution   **135**
  5.1   Standard Form 136
  5.2   State Density Function 146
  5.3   Asymptotic Free Energy 152
  5.4   Renormalized Posterior Distribution 154
  5.5   Conditionally Independent Case 162
  5.6   Problems 171

6 General Posterior Distribution   **177**
  6.1   Bayesian Decomposition 177
  6.2   Resolution of Singularities 181
  6.3   General Asymptotic Theory 190
  6.4   Maximum A Posteriori Method 196
  6.5   Problems 203

7 Markov Chain Monte Carlo   **207**
  7.1   Metropolis Method 207
      7.1.1   Basic Metropolis Method 209
      7.1.2   Hamiltonian Monte Carlo 211
      7.1.3   Parallel Tempering 215
  7.2   Gibbs Sampler 217
      7.2.1   Gibbs Sampler for Normal Mixture 218
      7.2.2   Nonparametric Bayesian Sampler 221
  7.3   Numerical Approximation of Observables 225
      7.3.1   Generalization and Cross Validation Losses 225
      7.3.2   Numerical Free Energy 226
  7.4   Problems 229

8 Information Criteria   **231**
  8.1   Model Selection 231
      8.1.1   Criteria for Generalization Loss 232
      8.1.2   Comparison of ISCV with WAIC 240

        8.1.3   Criteria for Free Energy 245
        8.1.4   Discussion for Model Selection 250
    8.2 Hyperparameter Optimization 251
        8.2.1   Criteria for Generalization Loss 253
        8.2.2   Criterion for Free Energy 257
        8.2.3   Discussion for Hyperparameter Optimization 259
    8.3 Problems 264

9 **Topics in Bayesian Statistics**   **267**
    9.1 Formal Optimality 267
    9.2 Bayesian Hypothesis Test 270
    9.3 Bayesian Model Comparison 275
    9.4 Phase Transition 277
    9.5 Discovery Process 282
    9.6 Hierarchical Bayes 286
    9.7 Problems 291

10 **Basic Probability Theory**   **293**
    10.1 Delta Function 293
    10.2 Kullback-Leibler Distance 294
    10.3 Probability Space 296
    10.4 Empirical Process 302
    10.5 Convergence of Expected Values 303
    10.6 Mixture by Dirichlet Process 306

References   **309**

Index   **317**

Preface

The purpose of this book is to establish a mathematical theory of Bayesian statistics.

In practical applications of Bayesian statistical inference, we need to prepare a statistical model and a prior for a given sample, then estimate the unknown true distribution. One of the most important problems is devising a method how to construct a pair of a statistical model and a prior, although we do not know the true distribution. The answer based on mathematical theory to this problem is given by the following procedures.

(1) Firstly, we construct the universal and mathematical laws between Bayesian observables which hold for an arbitrary triple of a true distribution, a statistical model, and a prior.

(2) Secondly, by using such laws, we can evaluate how appropriate a set of a statistical model and a prior is for the unknown true distribution.

(3) And lastly, the most suitable pair of the statistial model and the prior is employed.

The conventional approach to such a purpose has been based on the assumption that the posterior distribution can be approximated by some normal distribution. However, the new statistical theory introduced by this book holds for arbitrary posterior distribution, demonstrating that the application field will be extended. The author expects that also new statistical methodology which enables us to manupulate complex and hierarchical statistical models such as normal mixtures or hierarhical neural networks will be based on the new mathematical theory.

Sumio Watanabe

Chapter 1

Definition of Bayesian Statistics

In the first chaper, we introduce basic concepts in Bayesian statistics. In this book, we assume that there exists an unknown true distribution or unknown information source from which random variables are generated. Also we assume that an arbitrary set of a statistical model and a prior is prepared by a statistician who does not know the true distribution. Hence, from the mathematical point of view, the theory proposed in this book holds for an arbitrary set of a true distribution, a statistical model, and a prior.

The contents of this chaper include:

(1) In statistical estimation, we prepare a statistical model and a prior, though a true distribution is unknown. Hence evaluation of a statistical model and a prior is necessary.

(2) Several examples of probability distributions are introduced.

(3) It is assumed that an information source is represented by a probability distribution, which is called a true distribution.

(4) The posterior and the predicitive distributions are defined for a given statistical model and a prior.

(5) Two examples of posterior distributions are illustrated. In a simple estimation problem, the posterior distribution can be approximated by a normal distribution, whereas in a complex or hierarchical model, the result is far from any normal distribution. In this book we establish Bayesian theory which holds for both cases.

(6) The generalization loss is estimated by the cross validation loss and the widely applicable information criterion (WAIC).

(7) The marginal likelihood and the free energy of statistical estimation are

introduced.

(8) Statistical estimation in a conditional independent case is studied, in which the cross validation loss can not be used but WAIC can be.

For readers who are new to probability theory, chapter 10 will be helpful.

1.1   Bayesian Statistics

In this book, we assume that a sample $\{x_1, x_2, ..., x_n\}$ is taken from some true probability distribution $q(x)$,

$$\{x_1, x_2, ..., x_n\} \sim q(x).$$

This process is represented by a conditional probability distribution of a sample $\{x_1, x_2, ..., x_n\}$ for a given true distribution $q(x)$,

$$P(x_1, x_2, ..., x_n | q).$$

If we knew both $P(x_1, x_2, ..., x_n | q)$ and $P(q)$, where $P(q)$ is an *a priori* probability distribution of a true distribution, then by Bayes' theorem,

$$P(q | x_1, x_2, ..., x_n) = \frac{P(x_1, x_2, ..., x_n | q) P(q)}{\sum_q P(x_1, x_2, ..., x_n | q) P(q)},$$

which would give the statistical inference of $q(x)$ from a sample $\{x_1, x_2, ..., x_n\}$. However, in the real world, we do not have any information about either of them, showing that $P(q | x_1, x_2, ..., x_n)$ cannot be obtained.

A problem whose answer cannot be uniquely determined because of the lack of the information is called an ill-posed problem. Statistical inferences in the real world are ill-posed. In an ill-posed problem, we cannot determine a uniquely optimal method by which a correct answer is automatically obtained, which leads us to propose a new way:

Choose method → Result → Evaluate chosen method.

It might seem that such an evaluation is impossible because we do not have any information about the true distribution. However, in Bayesian statistics, there are mathematical laws which hold for an arbitrary set of a true distribution, a statistical model, and a prior. By using formulas derived from the mathematical laws, we can evaluate the appropriateness of the set of a statistical model and a prior even if a true distribution is unknown. The purpose of this book is to establish such mathematical laws.

In Bayesian statistics, we set a statistical model $p(x|w)$ and a prior $\varphi(w)$, where $p(x|w)$ is a conditional probability density of x for a given parameter w and $\varphi(w)$ is a probability density of w. Both $p(x|w)$ and $\varphi(w)$ are prepared by a statistician who does not know a true distribution. Hence they may be quite different from or inappropriate for the true distribution. If a sample $\{x_1, x_2, ..., x_n\}$ consists of independent sample points from a true distribution $q(x)$, then the posterior probability density function of w is defined by

$$p(w|x_1, x_2, ..., x_n) = \frac{\displaystyle\prod_{i=1}^{n} p(x_i|w)\varphi(w)}{\displaystyle\int \prod_{i=1}^{n} p(x_i|w)\varphi(w)dw}.$$

This is the definition of the posterior distribution. The estimated probability density function of x is defined by

$$\hat{p}(x) = \int p(x|w)p(w|x_1, x_2, ..., x_n)dw.$$

This is also the definition of the predictive distribution. A statistician estimates unknown $q(x)$ by $\hat{p}(x)$. For an arbitrary triple $(q(x), p(x|w), \varphi(w))$, we can define the Bayesian inference by this procedure, however, we need to examine whether a statistical model and a prior are appropriate for the unknown true distribution. Figure 1.1 shows the process of Bayesian estimation.

Remark 1. If we knew the true prior $\varphi_0(w)$ to which a parameter as a random variable is subject, and if a sample $\{x_1, x_2, ..., x_n\}$ was independently taken from the true conditional probability density $p_0(x|w)$, then the predictive distribution $\hat{p}(x)$ using $p_0(x|w)$ and $\varphi_0(w)$ would be the uniquely best inference. This is called the formal optimality of the Bayesian inference. See Section 9.1. However, in the real world, we do not know either $p_0(x|w)$ or $\varphi_0(x)$, indicating that we need evaluation because $\hat{p}(x)$ may be quite different from $q(x)$.

The candidate set $p(x|w)$ and $\varphi(w)$ is prepared by a statistician without any information about the true distribution. If the modeling $(p(x|w), \varphi(w))$ is appropriate for the unknown true $q(x)$, then it is expected that $\hat{p}(x) \approx q(x)$. However, if otherwise then $\hat{p}(x) \neq q(x)$. Hence we need a method to evaluate the appropriateness of the modeling $(p(x|w), \varphi(w))$ without any information about $q(x)$. In this book, we show such a method can be made based on mathematical laws which hold for arbitrary $(q(x), p(x|w), \varphi(w))$.

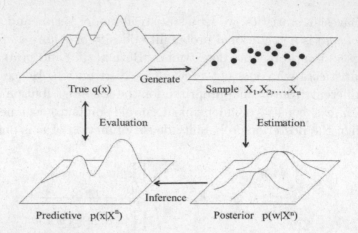

Figure 1.1: Framework of Bayesian inference. The procedure of Bayesian estimation is shown. A sample X^n is taken from unknown true distribution $q(x)$. A statistician sets a statistical model and a prior, then the posterior density $p(w|X^n)$ is obtained. The true distribution $q(x)$ is estimated by a predictive density $p(x|X^n)$, whose accuracy is evaluated by using mathematical laws.

1.2   Probability Distribution

Let us introduce a basic probability theory. For a reader who needs mathematical probability theory, Chapter 10 may be of help.

Let $x = (x_1, x_2, ..., x_N)$ be a vector contained in the N dimensional real Euclidean space $\mathbb{R}^N$. A real valued function

$$q(x) = q(x_1, x_2, ..., x_N)$$

is said to be a probability density function if it satisfies

- For arbitrary $x \in \mathbb{R}^N$, $q(x) \geq 0$,

- $\int q(x)dx = \int \int \cdots \int q(x_1, x_2, ..., x_N)dx_1 dx_2 \cdots dx_N = 1$.

Let A be a subset of $\mathbb{R}^N$ which has an finite integral value

$$\int_A q(x)dx.$$

A function Q of a set A defined by

$$Q(A) \equiv \int_A q(x)dx$$

is called a probability distribution. Note that $Q(\mathbb{R}^N) = 1$ and $Q(\varnothing) = 0$, where $\varnothing$ is the empty set.

If the probability that a variable X is in a set A is equal to $Q(A)$, then X is called a random variable and $q(x)$ and Q are called the probability density function and the probability distribution of a random variable X, respectively. Also it is said that a random variable X is subject to a probability density $q(x)$ or a probability distribution Q.

Example 1. (1) Let $N = 1$. A probability density function of a uniform distribution on $[a_0, b_0]$ $(a_0 < b_0)$ is given by

$$q(x) = \begin{cases} \frac{1}{b_0 - a_0} & (a_0 < x \le b_0) \\ 0 & (\text{otherwise}) \end{cases}.$$

(2) Let S be an $N \times N$ positive definite matrix and $m \in \mathbb{R}^N$. A normal distribution which has an average m and a covariance S is defined by

$$q(x) = \frac{1}{C} \exp\left(-\frac{1}{2}(x - m, S^{-1}(x - m))\right),$$

where $(\ ,\ )$ is the inner product in $\mathbb{R}^N$ and

$$C = (2\pi)^{N/2} \sqrt{\det(S)}.$$

(3) Let $H(x)$ be a function of $x \in \mathbb{R}^N$ and $\beta > 0$. If

$$Z(\beta) = \int \exp(-\beta H(x))dx$$

is finite, then a probability density function

$$q(x) = \frac{1}{Z(\beta)} \exp(-\beta H(x))$$

is called an equilibrium state of a Hamilton function $H(x)$ with the inverse temperature β.

The delta function $\delta(x)$ is characterized by two conditions,

$$\delta(x) = \begin{cases} +\infty & (\text{if } x = 0) \\ 0 & (\text{if } x \ne 0) \end{cases},$$

and

$$\int \delta(x)dx = 1.$$

The delta function can be understood as the probability density function of a random variable $X = 0$. Its probability distribution is given by

$$Q(A) = \left\{ \begin{array}{ll} 1 & (\text{if } 0 \in A) \\ 0 & (\text{if } 0 \notin A) \end{array} \right..$$

If a random variable X satisfies that $X = 1$ and $X = 2$ with probabilities $1/3$ and $2/3$ respectively, then its probability density function is

$$q(x) = \frac{1}{3}\delta(x - 1) + \frac{2}{3}\delta(x - 2).$$

If X is a random variable which is subject to $q(x)$ and Q, then $Y = f(X)$ is also a random variable which is subject to

$$p(y) = \int \delta(y - f(x))q(x)dx,$$

$$P(A) = \int_{f(x)\in A} q(x)dx.$$

These equations hold even if $f(x)$ is not one-to-one.

Remark 2. The function $\delta(x)$ is not an ordinary function of x. However, it is mathematically well-defined by Schwartz distribution theory and Sato hyperfunction theory. In this book, the delta function is necessary to study posterior distributions which cannot be approximated by any normal distribution.

Assume that a random variable X is subject to a probability density $q(x)$. The expected value, the average, or the mean of a random variable X on $\mathbb{R}^N$ is defined by

$$\mathbb{E}[X] = \int x\, q(x)dx,$$

if the right hand side is finite. The expected value of $Y = f(X)$ is

$$\mathbb{E}[Y] = \int y\, p(y)dy = \int y \int \delta(y - f(x))q(x)dxdy$$

$$= \int f(x)q(x)dx.$$

The covariance matrix of X is defined by

$$\begin{aligned} \mathbb{V}[X] &= \mathbb{E}[(X - \mathbb{E}[X])(X - \mathbb{E}[X])^T] \\ &= \mathbb{E}[XX^T] - \mathbb{E}[X]\mathbb{E}[X^T], \end{aligned}$$

if the right hand side is finite, where $(\ )^T$ shows the transposed matrix. If $N = 1$, then $\mathbb{V}[X]$ and $\mathbb{V}[X]^{1/2}$ are called the variance and the standard deviation, respectively.

Let (X, Y) be a pair of random variables which is subject to a probability density $q(x, y)$ on $\mathbb{R}^M \times \mathbb{R}^N$. Here $q(x, y)$ is called a simultaneous probability density of (X, Y). Then X and Y are subject to the probability densities

$$\begin{aligned} q(x) &= \int q(x, y)dy, \\ q(y) &= \int q(x, y)dx, \end{aligned}$$

where $q(x)$ and $q(y)$ are called marginal probability densities of X and Y, respectively. The conditional probability density of Y for a given X is defined by

$$q(y|x) = \frac{q(x, y)}{q(x)}.$$

If $q(x) = 0$, then $q(y|x)$ is not defined, however, we define $0 \cdot q(y|x) = 0$. The conditional probability density function $q(x|y)$ is also defined by $q(x, y)/q(y)$. Then it follows that

$$q(x, y) = q(y|x)q(x) = q(x|y)q(y).$$

This equation is sometimes referred to as Bayes' theorem.

1.3 True Distribution

In this book, it is mainly assumed that a sample is a set of random variables taken from a true distribution.

Let n be a positive integer. A set of $\mathbb{R}^N$-valued random variables X_1, X_2, ..., X_n is sometimes denoted by

$$X^n = (X_1, X_2, ..., X_n).$$

Throughout this book, the notation n is used for the number of random variables. Sometimes X^n and n are referred to as a sample and a sample size, respectively. A realized value of X^n in a trial is denoted by

$$x^n = (x_1, x_2, ..., x_n).$$

If X^n is subject to a probability density function,

$$q(x_1)q(x_2) \cdots q(x_n)$$

then X^n is called a set of independent random variables which are subject to the same probability density $q(x)$. Here $q(x)$ is sometimes referred to as a true probability density. In the practical applications, we do not know $q(x)$, but we assume there exists such a density $q(x)$.

For an arbitrary function $f : x^n \mapsto f(x^n) \in \mathbb{R}$, the expected value of $f(X^n)$ over X^n is denoted by $\mathbb{E}[\ ]$. That is to say,

$$\mathbb{E}[f(X^n)] = \int \int \cdots \int f(x^n) \prod_{i=1}^{n} q(x_i)dx_1 dx_2 \cdots dx_n.$$

The variance of $f(X^n)$ is denoted by

$$\mathbb{V}[f(X^n)] = \mathbb{E}[f(X^n)^2] - \mathbb{E}[f(X^n)]^2.$$

The average and empirical entropies of the true distribution are respectively defined by

$$S = -\int q(x) \log q(x) dx, \tag{1.1}$$

$$S_n = -\frac{1}{n} \sum_{i=1}^{n} \log q(X_i). \tag{1.2}$$

Then by the definition,

$$\mathbb{E}[S_n] = S, \tag{1.3}$$

$$\mathbb{V}[S_n] = \frac{1}{n}\left[\int q(x)(\log q(x))^2 dx - S^2 \right]. \tag{1.4}$$

Remark 3. (The number n) In statistics, x_i and x^n are referred to as a sample point and a sample, respectively. The number n is called a sample size. In machine learning, x_i and x^n are referred to as a datum and a set of training data. The number n is called the number of training data or the number of examples. In this book, the notation n is used for the number of random variables, which is equal to the sample size in statistics and the number of training data in machine learning.

If a set of $\mathbb{R}^M \times \mathbb{R}^N$-valued random variables (X^n, Y^n) is subject to a probability density function,

$$q(x_1, y_1)q(x_2, y_2) \cdots q(x_n, y_n),$$

then (X^n, Y^n) is a set of independent random variables. The average and empirical entropies are repsectively given by

$$S = -\int q(x, y) \log q(x, y) dx dy, \tag{1.5}$$

$$S_n = -\frac{1}{n} \sum_{i=1}^{n} \log q(X_i, Y_i). \tag{1.6}$$

These are referred to as the simlutaneous average and empirical entropies. We often need to estimate the conditional probability density function $q(y|x)$ under the condition that (X^n, Y^n) is obtained. The average and the empirical entropies of the true conditional distribution are respectively defined by

$$S = -\int q(x, y) \log q(y|x) dx dy, \tag{1.7}$$

$$S_n = -\frac{1}{n} \sum_{i=1}^{n} \log q(Y_i|X_i). \tag{1.8}$$

Then by the definition,

$$\mathbb{E}[S_n] = S,$$
$$\mathbb{V}[S_n] = \frac{1}{n}\left[\int q(x, y)(\log q(y|x))^2 dx dy - S^2\right].$$

Sometimes we need to study cases when (X^n, Y^n) is not independent, but Y^n for a given X^n is independent. Such a case is explained in Sections 1.8 and 5.5.

1.4 Model, Prior, and Posterior

Let W be a set of parameters which is a subset of d dimensional real Euclidean space $\mathbb{R}^d$. A statistical model or a learning machine is defined by $p(x|w)$ which is a conditional probability density of $x \in \mathbb{R}^N$ for a given parameter $w \in W$. A prior $\varphi(w)$ is a probability density of $w \in W$.

Let $X^n = (X_1, X_2, ..., X_n)$ be a set of random variables which are independently subject to a probability density function $q(x)$. For an arbitrary pair $(p(x|w), \varphi(w))$, the posterior probability density is defined by

$$p(w|X^n) = \frac{1}{Z(X^n)} \varphi(w) \prod_{i=1}^{n} p(X_i|w),$$

where $Z(X^n)$ is defined by

$$Z(X^n) = \int \varphi(w) \prod_{i=1}^{n} p(X_i|w) dw,$$

which is called the partition function or the marginal likelihood.   The posterior average or the expected value over the posterior distribution is denoted by $\mathbb{E}_w[\ ]$. For an arbitrary function $f(w)$,

$$\mathbb{E}_w[f(w)] = \int f(w) p(w|X^n) dw$$

$$= \frac{1}{Z(X^n)} \int f(w) \varphi(w) \prod_{i=1}^{n} p(X_i|w) dw.$$

The posterior variance is also defined by

$$\mathbb{V}_w[f(w)] = \mathbb{E}_w[f(w)^2] - \mathbb{E}_w[f(w)]^2.$$

Remark that the expectation operator $\mathbb{E}_w[\ ]$ depends on a set X^n, hence $\mathbb{E}_w[f(w)]$ is not a constant but a random variable. The predictive density function is defined by

$$p(x|X^n) = \mathbb{E}_w[p(x|w)] = \int p(x|w) p(w|X^n) dw.$$

For a given sample X^n, the Bayesian estimation of the true distribution is defined by $p(x|X^n)$.

Remark 4. Sometimes a prior function which satisfies

$$\int \varphi(w) dw = \infty$$

is employed. If $\int \varphi(w) dw < \infty$, it is called proper, because it is normalized so that $\int \varphi(w) dw = 1$. If $\varphi(w)$ is not proper, then it is called improper. Even for an improper prior, the posterior and predictive probability densities can be defined by the same equation if $Z(X^n)$ $(n \geq 1)$ is finite and they are well-defined.

Let (X^n, Y^n) be a set of random variables which are independently subject to a probability density $q(x, y) = q(y|x)q(x)$. For an arbitrary pair $(p(y|x, w), \varphi(w))$, the posterior probability density is defined by

$$p(w|X^n, Y^n) = \frac{1}{Z(X^n, Y^n)} \varphi(w) \prod_{i=1}^{n} p(Y_i|X_i, w), \qquad (1.9)$$

where $Z(X^n, Y^n)$ is defined by

$$Z(X^n, Y^n) = \int \varphi(w) \prod_{i=1}^{n} p(Y_i|X_i, w) dw.$$

The posterior average is also denoted by $\mathbb{E}_w[\ ]$. For an arbitrary function $f(w)$,

$$\begin{aligned} \mathbb{E}_w[f(w)] &= \int f(w) p(w|X^n, Y^n) dw \\ &= \frac{1}{Z(X^n, Y^n)} \int f(w) \varphi(w) \prod_{i=1}^{n} p(Y_i|X_i, w) dw. \end{aligned}$$

The posterior variance is also defined by

$$\mathbb{V}_w[f(w)] = \mathbb{E}_w[f(w)^2] - \mathbb{E}_w[f(w)]^2.$$

The predictive density function is defined by

$$p(y|x, X^n, Y^n) = \mathbb{E}_w[p(y|x, w)] = \int p(y|x, w) p(w|X^n, Y^n) dw. \qquad (1.10)$$

For the case when Y^n is independent for a given X^n, see Sections 1.8 and 5.5.

1.5   Examples of Posterior Distributions

Let us illustrate several posterior distributions. In a simple statistical model, the posterior distribution can be approximated by a normal distribution, but not in a complex model. One of the main purposes of this book is to establish the universal mathematical theory which holds in both cases.

Example 2. (Normal Distribution) A normal distribution whose average and standard deviation are (a, σ) is defined by

$$p(x|a, \sigma) = \frac{1}{\sqrt{2\pi\sigma^2}} \exp\left(-\frac{(x-a)^2}{2\sigma^2}\right). \qquad (1.11)$$

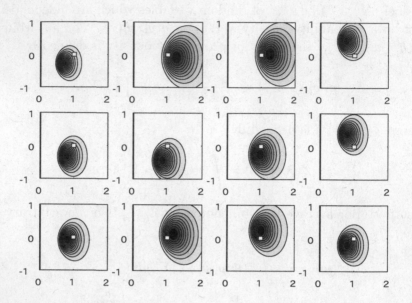

Figure 1.2: Posterior distributions of a statistical model eq.(1.11) with $n = 10$ are shown. The white square is the true parameter. Even for an identical true probability density, posterior distributions fluctuate depending on samples.

Let us study a case when a prior is set

$$\varphi(a, \sigma) = \left\{ \begin{array}{ll} 1/4 & (|a| < 1,\ 0 < \sigma < 2) \\ 0 & \text{(otherwise)} \end{array} \right. .$$

Assume that a true distribution is $q(x) = p(x|0, 1)$ and the number of independent random variables is n. In this case, the parameter that attains the true density is unique,

$$q(x) = p(x|a, \sigma) \Longleftrightarrow (a, \sigma) = (1, 0).$$

In Figure 1.5, posterior distributions for 12 different samples with $n = 10$ are shown using gray scale. The white square is the position of the true parameter $(0, 1)$. Even if a true probability density function is identical, the posterior distribution has fluctuations according to a sample.

In Figure 1.3, posterior distributions for 12 different samples with $n = 50$ are shown. In this case, the posterior distributions concentrate in a

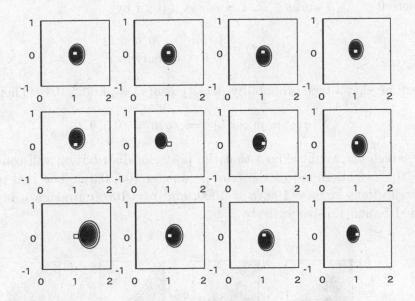

Figure 1.3: Posterior distributions of a statistical model eq.(1.11) with $n = 50$. The white square is the true parameter. Posterior distributions can be approximated by some normal distribution.

neighborhood of the true parameter when n becomes large. It seems that the posterior distribution can be approximated by some normal distribution, hence we expect that a conventional statistical theory using the posterior normality can be applied to evaluation of statistical modeling.

Example 3. (Normal Mixture) One might think that a posterior distribution can be approximated by some normal distribution if d/n is sufficiently small, where n and d are the number of random variables and the dimension of the parameter, respectively. However, such consideration often fails even in unspecial statistical models. Let $N(x)$ be the standard normal distribution,

$$N(x) = \frac{1}{\sqrt{2\pi}} \exp\left(-\frac{x^2}{2}\right).$$

A normal mixture which has a parameter (a, b) is defined by

$$p(x|a, b) = (1 - a)N(x) + aN(x - b), \tag{1.12}$$

where $0 \leq a \leq 1$ and $b \in \mathbb{R}$. Let us set a prior by

$$\varphi(a, \sigma) = \left\{ \begin{array}{ll} 1 & (0 < a, b < 1) \\ 0 & \text{(otherwise)} \end{array} \right. .$$

Assume that a true probability density is $q(x) = p(x|0.5, 0.3)$. Then

$$q(x) = p(x|a, b) \iff (a, b) = (0.5, 0.3),$$

by which one might expect that the posterior distribution will concentrate on the neighborhood of the true parameter $(0.5, 0.3)$. The real posterior distributions for $n = 100$, $n = 1000$, and $n = 10000$ are shown in Figures 1.4, 1.5, and 1.6, respectively.

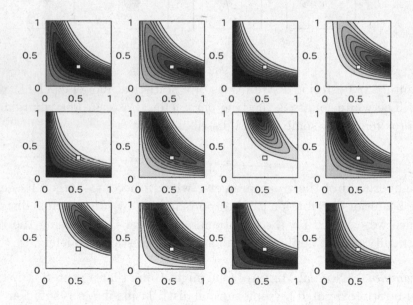

Figure 1.4: Postrior distributions of a statistical model eq.(1.12) with $n = 100$. The white square is the true parameter. The posterior distributions are far from any normal distribution and their fluctuations are very large. The statistical theory in this book enables us to estimate the generalization loss even in this case.

Even when $n = 10000$, the posterior distribution cannot be approximated by any normal distribution. The regular statistical theory, which

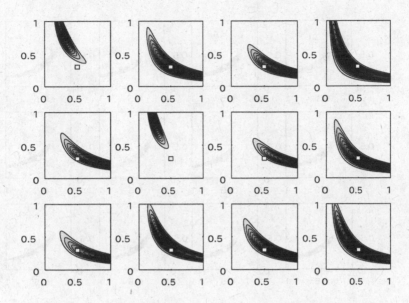

Figure 1.5: Postrior distributions of a statistical model eq.(1.12) with $n = 1000$. The white square is the truc parameter. The number $n = 1000$ seems to be sufficiently large, however, the posterior distributions are far from any normal distribution and their fluctuations are very large.

assumes that the posterior distribution can be approximated by a normal distribution, cannot be applied to this case. Therefore, in order to use the regular asymptotic theory, the condition $n >> 10000$ is necessary, if otherwise it has been difficult to establish a statistical hypothesis test or a statistical model selection. In this book, we show a new statistical theory which holds even if $n = 100$ can be established by a mathematical base.

Both statistical models given by eq.(1.11) and eq.(1.12) are employed in many statistical inferences. In the former model, $p(x|m, s)$ represents one normal distribution for an arbitrary (m, s), whereas in the latter model, $p(x|a, b)$ represents one or two normal distributions, depending on the parameter. In fact, if $ab = 0$, then $p(x|a, b)$ is a standard normal distribution. Hence the parameter (a, b) is not a simple parameter but affects the structure of a statistical model. In general, if a statistical model has hierarchical structure or a hidden variable such as the latter model, then the posterior distribution cannot be approximated by a normal distribution in general.

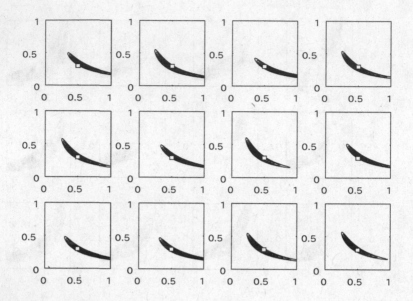

Figure 1.6: Postrior distributions of a statistical model eq.(1.12) with $n =$ 10000. The white square is the true parameter. In this case, $n = 10000$ and the number of parameters is 2, however, the posterior distributions can not be approximated by any normal distribution.

In this book, in Chapter 4, we study the former statistical models, and in Chapters 5 and 6, we derive a new mathematical theory for both models.

From the mathematical point of view, the statistical model of eq.(1.11) does not have singularities, whereas that of eq.(1.12) does. The true parameter $(0.5, 0.3)$ in eq.(1.12) is a nonsingular point but lies near singularity $(0,0)$. In fact, the function from a parameter to a statistical model is not one-to-one,

$$\{(a, b); p(x|a, b) = p(x|0, 0)\} = \{(a, b); ab = 0\},$$

and $(a, b) = (0, 0)$ is a singularity of this set. It should be empasized that a singularity affects the posterior distribution even if the true parameter is not a singularity. Moereover, several statistical models used in infromation processing such as artificial neural networks and mixture models have many singularities. In general, if a statistical model has singularities, then the Bayesian estimation has better generalization performance than the maximum likelihood method, hence we need new statistical theory for the purpose of constructing hypothesis testing, model selection, and hyperparameter optimization for such singular statistical models.

1.6 Estimation and Generalization

In order to evaluate how accurate the predictive density is, we need an objective measure which indicates the difference between the true and the estimated probability density.

Definition 1. (Generalization and Training Losses) Let X^n be a sample which is independently taken from a true distribution $q(x)$ and $p(x|X^n)$ be a predictive density using a statistical model $p(x|w)$ and a prior $\varphi(w)$. The training and generalization losses are respectively defined by

$$T_n = -\frac{1}{n}\sum_{i=1}^{n}\log p(X_i|X^n), \tag{1.13}$$

$$G_n = -\int q(x)\log p(x|X^n)dx. \tag{1.14}$$

Note that both G_n and T_n are random variables. Let S be the entropy of a true distribution given by eq.(1.1). Then it immediately follows that

$$
\begin{aligned}
G_n - S &= -\int q(x)\log p(x|X^n)dx + \int q(x)\log q(x)dx \\
&= \int q(x)\log\frac{q(x)}{p(x|X^n)}dx \\
&= K(q(x)\|p(x|X^n)),
\end{aligned} \tag{1.15}
$$

where $K(q(x)\|p(x|X^n))$ is the Kullback-Leibler distance from $q(x)$ to $p(x|X^n)$. For the definition of Kullback-Leibler distance, see Section 10.2. In general,
(1) $K(q(x)\|p(x|X^n)) \geq 0$.
(2) $K(q(x)\|p(x|X^n)) = 0$ if and only if $K(q(x)\|p(x|X^n)) = 0$.
Hence
(1) $G_n \geq S$.
(2) $G_n - S = 0$ if and only if $q(x) = p(x|X^n)$.
That is to say, the smaller G_n is, the more precise estimation is obtained according to Kullback-Leibler distance. The random variables $G_n - S$ and $T_n - S_n$ are called generalization and training errors respectively.

Remark 5. Assume that we have two sets of statistical models and priors,

$$(p_1(x|w), \varphi_1(w)), \quad (p_2(x|w), \varphi_2(w)).$$

Let $p_1(x|X^n)$ and $p_2(x|X^n)$ be predictive densities of two pairs respectively, and $G_n(1)$ and $G_n(2)$ be their generalization losses. Since the entropy S does not depend on either a model or a prior,

$$G_n(1) > G_n(2) \iff K(q(x)\|p_1(x|X^n)) > K(q(x)\|p_2(x|X^n)),$$

which shows that the smaller generalization loss is equivalent to the smaller Kullback-Leibler distance. Two training losses $T_n(1)$ and $T_n(2)$ can be defined for both sets, but they do not have such properties. In other words, the smaller training loss does not mean a smaller generalization error.

Definition 2. Assume $n \geq 2$. Let $X^n \setminus X_i$ be a set of random variables X_1, X_2, ..., X_n which does not contain X_i and $p(x|X^n \setminus X_i)$ be the predictive density using $X^n \setminus X_i$. The cross validation loss is defined by

$$C_n = -\frac{1}{n} \sum_{i=1}^{n} \log p(X_i|X^n \setminus X_i). \tag{1.16}$$

Also $C_n - S_n$ is called a cross validation error.

Remark 6. The definition eq.(1.16) is called the leave-one-out cross validation loss. There are several kinds of cross validation losses, however, we mainly study the leave-one-out one in this book, because it is most accuate as an estimator of the generalization loss. The cross validation loss can be defined even if X^n is dependent. However, if X^n is dependent, then it is not an appropriate estimator of the generalization loss. In fact, the following theorem needs independence.

Theorem 1. *Assume that X^n is independent. Then the following holds.*
(1) Assume that the expectation values of G_n and C_n are finite. Then

$$\mathbb{E}[C_n] = \mathbb{E}[G_{n-1}]. \tag{1.17}$$

(2) The cross validation loss satisfies the relation,

$$C_n = \frac{1}{n} \sum_{i=1}^{n} \log \mathbb{E}_w \left[\frac{1}{p(X_i|w)} \right].$$

(3) For an arbitrary set of random variables X^n,

$$C_n \geq T_n.$$

The equality $C_n = T_n$ holds if and only if $p(X_i|w)$ ($i = 1, 2, ..., n$) is a constant function of w on $\{w \in W; p(w|X^n) > 0\}$.

Proof. (1) The set $X^n \setminus X_i$ does not contain X_i, hence

$$
\begin{aligned}
\mathbb{E}[C_n] &= -\mathbb{E}\Big[\frac{1}{n}\sum_{i=1}^{n}\log p(X_i|X^n \setminus X_i)\Big] \\
&= -\frac{1}{n}\sum_{i=1}^{n}\mathbb{E}\Big[\int q(x)\log p(x|X^n \setminus X_i)dx\Big] \\
&= \mathbb{E}[G_{n-1}].
\end{aligned}
$$

(2) For an arbitrary i,

$$
\prod_{j=1}^{n} p(X_j|w) = p(X_i|w)\prod_{j\neq i} p(X_j|w).
$$

By the definition of the cross validation loss,

$$
\begin{aligned}
C_n &= -\frac{1}{n}\sum_{i=1}^{n}\log \frac{\int p(X_i|w)\varphi(w)\prod_{j\neq i} p(X_j|w)dw}{\int \varphi(w)\prod_{j\neq i} p(X_i|w)dw} \\
&= \frac{1}{n}\sum_{i=1}^{n}\log \frac{\int (1/p(X_i|w))\varphi(w)\prod_{j=1}^{n} p(X_i|w)dw}{\int \varphi(w)\prod_{j=1}^{n} p(X_i|w)dw},
\end{aligned}
$$

which shows (2) of the lemma.
(3) By using the result (2),

$$
C_n - T_n = \frac{1}{n}\sum_{i=1}^{n}\log\Big(\mathbb{E}_w[p(X_i|w)]\mathbb{E}_w[1/p(X_i|w)]\Big).
$$

From Cauchy-Schwarz inequality, it follows that

$$
\mathbb{E}_w[p(X_i|w)]\mathbb{E}_w[1/p(X_i|w)] \geq \mathbb{E}_w[p(X_i|w)^{1/2}p(X_i|w)^{-1/2}]^2 = 1.
$$

The equality $C_n = T_n$ holds if and only if $p(X_i|w)^{1/2} \propto p(X_i|w)^{-1/2}$ as a function of w, which concludes (3). $\qquad\square$

Remark 7. (1) The conditional probability density $p(X_i|X^n \setminus X_i)$ is the predictive density of X_i based on a sample X^n leaving X_i out. Thus the average cross validation loss is naturally an unbiased estimator of the generalization loss for $n-1$. In the real world, the generalization loss cannot be calculated because we do not know the true distribution $q(x)$, whereas the cross validation loss can be obtained using only a sample X^n. There are two issues about the cross validation.

- Although the averages of C_n and G_{n-1} are equal to each other, their variances are not directly derived from their definitions. In the following chapter, we prove that the standard deviations $G_n - S$ and $C_n - S_n$ are asymptotically equal to each other, in proportion to $1/n$.

- If the average by the posterior distribution is numerically approximated, then

$$\text{ISCV} \;=\; \frac{1}{n}\sum_{i=1}^{n} \log \mathbb{E}_w\!\left[\frac{1}{p(X_i|w)}\right],$$

is called the importance sampling cross validation loss. Note that the approximated cross validation loss

$$\text{CV} \;=\; -\frac{1}{n}\sum_{i=1}^{n} \log \mathbb{E}_w^{(-i)}[p(X_i|w)],$$

where $\mathbb{E}_w^{(-i)}[\ ]$ shows the posterior average for $X^n \backslash X_i$, is different from the importance sampling cross validation loss if the posterior density is not precisely approximated.

Definition 3. Assume $n \geq 1$. Let X^n be a set of random variables. The widely applicable information criterion (WAIC) is defined by

$$W_n = T_n + \frac{1}{n}\sum_{i=1}^{n} \mathbb{V}_w[\log p(X_i|w)], \qquad (1.18)$$

where $\mathbb{V}_w[\ ]$ shows the posterior variance. Also $W_n - S_n$ is called a WAIC error.

In the following chapters, we show that, if X^n is independent, then WAIC is asymptotically equivalent to the cross validation loss,

$$W_n = C_n + O_p(1/n^2), \qquad (1.19)$$

and

$$\mathbb{E}[W_n] = \mathbb{E}[C_n] + O(1/n^2). \qquad (1.20)$$

Moreover, there are several cases even if X^n is dependent,

$$\mathbb{E}[W_n] = \mathbb{E}[G_n] + o(1/n). \qquad (1.21)$$

For example, the formula $\mathbb{E}[C_n] = \mathbb{E}[G_{n-1}]$ does not hold in conditional independent cases such as regression problems of fixed inputs or time series prediction, whereas $\mathbb{E}[W_n] = \mathbb{E}[G_n] + o(1/n)$ holds even for such cases.

Remark 8. The cross validation loss and WAIC can be employed for evaluation of a statistical model and a prior even if a prior is improper.

Remark 9. (Loss and error) The generalization, cross validation, and WAIC errors are defined by

$$G_n - S, \quad C_n - S_n, \quad W_n - S_n,$$

where S and S_n are the average and empirical entropies of a true distribution, respectively. In practical applications, we do not know the true distribution, resulting that S and S_n are unknown. However, neither S nor S_n depends on a statistical model and a prior. In model selection and hyperparameter optimization, minimizing losses are equivalent to minimizing errors. On the other hand, errors have smaller variances than losses, hence in numerical experiments, we often compare errors instead of losses.

1.7 Marginal Likelihood or Partition Function

If a prior $\varphi(w)$ satisfies $\int \varphi(w)dw = 1$, then the marginal likelihood or the partition function $Z(X^n) = Z(X_1, X_2, ..., X_n)$ satisfies

$$\int Z(x_1, x_2, ..., x_n)dx_1 dx_2 \cdots dx_n$$

$$= \int dw \, \varphi(w) \int dx_1 dx_2 \cdots dx_n \prod_{i=1}^{n} p(x_i|w) = 1.$$

Therefore $Z(x^n)$ can be understood as an estimated probability density function of X^n by using a statistical model $p(x|w)$ and a prior $\varphi(w)$. Therefore $Z(x^n)$ is sometimes written as $p(x^n)$.

The free energy or the minus log marginal likelihood is defined by

$$F_n = -\log Z(X^n). \tag{1.22}$$

Then by using notations $q(x^n) = \prod_{i=1}^{n} q(x_i)$ and $p(x^n) = Z(x^n)$,

$$\mathbb{E}[F_n] - nS = \int q(x^n) \log \frac{q(x^n)}{p(x^n)} dx^n,$$

which shows that $\mathbb{E}[F_n] - nS$ is equal to the Kullback-Leibler distance from the true density $q(x^n)$ to the estimated density $p(x^n)$. The smaller $\mathbb{E}[F_n]$ is equivalent to the smaller Kullback-Leibler distance between them. Note that $\mathbb{E}[G_n] - S$ is the average Kullback-Leibler distance from $q(x)$ to $p(x|X^n)$, whereas $\mathbb{E}[F_n] - nS$ is their sum.

Theorem 2. *Let $n \geq 1$. The average generalization loss is equal to the increase of the free energy,*

$$\mathbb{E}[G_n] = \mathbb{E}[F_{n+1}] - \mathbb{E}[F_n]. \tag{1.23}$$

Therefore the average free energy is the sum of the generalization loss.

$$\mathbb{E}[F_n] = \sum_{i=1}^{n-1} \mathbb{E}[G_i] + \mathbb{E}[F_1]. \tag{1.24}$$

Proof. Let X_{n+1} be a random variable which is independent of X^n and subject to the same probability density function $q(x)$. Then for an arbitrary function $f(x)$,

$$\int q(x) f(x) dx = \mathbb{E}_{X_{n+1}}[f(X_{n+1})].$$

By using this notation,

$$
\begin{aligned}
G_n &= -\int q(x) \log p(x|X^n) dx \\
&= -\mathbb{E}_{X_{n+1}}[\log p(X_{n+1}|X^n)] \\
&= -\mathbb{E}_{X_{n+1}}\left[\log \frac{\displaystyle\int p(X_{n+1}|w)\varphi(w) \prod_{i=1}^{n} p(X_i|w) dw}{\displaystyle\int \varphi(w) \prod_{i=1}^{n} p(X_i|w) dw}\right] \\
&= -\mathbb{E}_{X_{n+1}}[\log Z(X^{n+1})] + \log Z(X^n).
\end{aligned}
$$

The expected values over X^n of this equation show eq.(1.23). Therefore,

$$
\begin{aligned}
\mathbb{E}[F_n] &= \mathbb{E}[G_{n-1}] + \mathbb{E}[F_{n-1}] \\
&= \mathbb{E}[G_{n-1}] + \mathbb{E}[G_{n-2}] + \mathbb{E}[F_{n-2}] \\
&= \sum_{i=1}^{n-1} \mathbb{E}[G_i] + \mathbb{E}[F_1],
\end{aligned}
$$

which shows eq.(1.24). $\square$

Remark 10. (Marginal likelihood and free energy) By the definition $F_n = -\log Z(X^n)$, the correspondence between the free energy and the marginal likelihood is one-to-one. Hence one of them is obtained, and the other can be easily derived. However, in general, the asymptotic order of the marginal likelihood as a random variable is not equal to its average, whereas that of the free energy is equal to its average. Therefore, in studying asymptotic statistics, the free energy is the more convenient random variable than the marginal likelihood. Let us illustrate this fact. A marginal likelihood ratio function $r(x^n)$ is defined by

$$r(x^n) = \frac{Z(x^n)}{q(x^n)}.$$

Let X^n be a set of random variables which are independently subject to $q(x)$. Then

$$\mathbb{E}[r(X^n)] = \int r(x^n)q(x^n)dx^n = 1.$$

Therefore the average of $r(X^n)$ is always equal to one. However,

$$r(X^n) \to 0 \quad \text{in probability.}$$

For example, for $x, a \in \mathbb{R}$, a statistical model and a prior are defined by

$$p(x|a) = \frac{1}{\sqrt{2\pi}} \exp(-\frac{1}{2}(x-a)^2),$$

$$\varphi(a) = \frac{1}{\sqrt{2\pi}} \exp(-\frac{a^2}{2}),$$

and a true distribution is set as $q(x) = p(x|0)$, then

$$r(X^n) = \sqrt{\frac{2\pi}{n+1}} \exp\left\{ \frac{n}{2(n+1)} \left(\frac{1}{\sqrt{n}} \sum_{i=1}^{n} X_i \right) \right\}.$$

Since $(1/\sqrt{n}) \sum_{i=1}^{n} X_i$ is a random variable which is subject to the standard normal distribution, $r(X^n) \to 0$ in probability. Therefore the order of $r(X^n)$ is not equal to its average. On the other hand, the order of the random variable $-\log r(X^n)$ is equal to its average, because

$$-\log r(X^n) = \frac{1}{2}\log(n+1) - \frac{1}{2}\log(2\pi) - \frac{n}{2(n+1)} \left(\frac{1}{\sqrt{n}} \sum_{i=1}^{n} X_i \right).$$

Remark 11. (Simultaneous prediction) Let us compare Bayesian estimation and the other estimation from the simultaneous prediction point of view. Let $X_1, X_2, ..., X_n, X_{n+1}, ..., X_{n+m}$ be independent random variables which are subject to the same distribution. The simultaneous estimation of

$$X^{n+m} \setminus X^n = (X_{n+1}, X_{n+2}, ..., X_{n+m})$$

for a given sample X^n is

$$p(X^{n+m} \setminus X^n | X^n) = \frac{Z(X^{n+m})}{Z(X^n)}.$$

Hence

$$p(X^{n+m} \setminus X^n | X^n) = \prod_{j=1}^{m} \left(\frac{Z(X^{n+j})}{Z(X^{n+j-1})} \right),$$

resulting that

$$-\log p(X^{n+m} \setminus X^n | X^n) = -\sum_{j=1}^{m} \log \left(\frac{Z(X^{n+j})}{Z(X^{n+j-1})} \right).$$

The average of this equation is given by

$$\mathbb{E}[G_n] + \mathbb{E}[G_{n+1}] + \cdots + \mathbb{E}[G_{n+m-1}]. \tag{1.25}$$

On the other hand, let $\hat{w}$ be an estimator such as the maximum likelihood or the maximum *a posteriori* method determined by X^n. Then the generalization loss of $X^{n+m} \setminus X^n$ for a given X^n is

$$-\sum_{j=1}^{m} \log p(X_{n+j} | \hat{w}),$$

whose expected value is

$$-m \times \mathbb{E}[\log p(X | \hat{w})]. \tag{1.26}$$

By eq.(1.25), it is shown that, in Bayesian estimation, the predicted sample point is automatically used, recursively. However, by eq.(1.26), in other methods, that is not the case. In ordinary cases, the average generalization loss $\mathbb{E}[G_n]$ is a decreasing function of n, hence, from the simultaneous prediction point of view, Bayesian estimation is better than the other methods. By the same reason, in the prediction of high dimensional X, it is expected that the Bayesian estimation has the better performance than the other methods.

Remark 12. (Predictive measure and marginal likelihood) The cross validation loss measures the predictive loss which is defined by Kullback-Leibler distance between $q(x)$ and $p(x|X^n)$, whereas the free energy indicates the cumulative loss which is defined by Kullback-Leibler distance between $q(X^n)$

and $p(X^n)$. Both measures are important but different in Bayesian statistics. If they are used as criteria for choosing the best model or the best hyperparameter, then the chosen model or hyperparameter is different according to the criteria. In Chapter 8, we study mathematical properties of both measures.

Remark 13. (Meaning of the marginal likelihood) Assume that $P_0(p, \varphi)$ is the prior distribution of a model $p(x|w)$ and a prior $\varphi(w)$. Then the probability density of X^n for a given (p, φ) is

$$P(X^n|p, \varphi) = \int \prod_{i=1}^{n} p(X_i|w)\varphi(w)dw = Z(X^n).$$

By Bayes' theorem, the posterior probability density of (p, φ) for a given sample X^n is

$$P(p, \varphi|X^n) = \frac{P(X^n|p, \varphi)P_0(p, \varphi)}{P(X^n)}.$$

Although $Z(X^n)$ is not strictly equivalent to the maximization of the posterior probability $P(p, \varphi|X^n)$, if n is sufficiently large, the maximization of $Z(X^n)$ becomes equivalent to the maximization of the posterior probability.

Remark 14. (Asymptotic expansions of free energy and generalization loss) Let $f(n) = \mathbb{E}[F_n]$ and $g(n) = \mathbb{E}[G_n]$. Assume that there exist constants $\{A_i\}$ and $\{B_i\}$ such that asymptotic expansions

$$f(n) = A_1 n + A_2\sqrt{n} + A_3 \log n + A_4 \log\log n + O(1), \qquad (1.27)$$

$$g(n) = B_1 + \frac{B_2}{2\sqrt{n}} + \frac{B_3}{n} + \frac{B_4}{n\log n} + o(\frac{1}{n\log n}), \qquad (1.28)$$

hold for $n \to \infty$. Then by eq.(1.23), $A_i = B_i$ ($i = 1, 2, 3, 4$). It is important that the constant order term of the free energy $f(n)$ does not affect the generalization loss $g(n)$. Sometimes minimization of the free energy changes the constant order term but does not minimize the generalization loss. See Chapter 8. Also note that, mathematically speaking, even if $f(n)$ has an asymptotic expansion, $g(n)$ may not have any asymptotic expansion, however, if $g(n)$ has an asymptotic expansion, then its coefficients are uniquely determined by the asymptotic expansion of $f(n)$.

1.8   Conditional Independent Cases

In several practical applications, we need to study cases when X^n is dependent. In this section, let us assume that X^n is dependent but Y^n is conditionally independent, in other words, Y^n is independent for a given X^n. If a set of $\mathbb{R}^N$-valued random variables $Y^n = (Y_1, Y_2, ..., Y_n)$ is subject to a probability density function,

$$q(y_1|x_1)q(y_2|x_2)\cdots q(y_n|x_n)$$

for some fixed $x^n = (x_1, x_2, ..., x_n)$, then Y^n is called a conditionally independent random variables subject to a conditional probability density function $\prod_{i=1}^{n} q(y_i|x_i)$. For an arbitrary function $f : (x^n, y^n) \mapsto f(x^n, y^n) \in \mathbb{R}$, the expected value of $f(x^n, Y^n)$ over Y^n is denoted by $\mathbb{E}[\ ]$. That is to say

$$\mathbb{E}[f(x^n, Y^n)] = \int\int \cdots \int f(x^n, y^n) \prod_{i=1}^{n} q(y_i|x_i)dy_1 dy_2 \cdots dy_n,$$

which is a function of x^n. The average and the empirical entropies of the true distribution are respectively defined by

$$S = -\frac{1}{n}\sum_{i=1}^{n}\int q(y|x_i)\log q(y|x_i)dy, \tag{1.29}$$

$$S_n = -\frac{1}{n}\sum_{i=1}^{n}\log q(Y_i|x_i). \tag{1.30}$$

Note that both posterior and predictive probability densities are given by the same equations as eq.(1.9) and eq.(1.10), respectively.

$$p(w|x^n, Y^n) = \frac{1}{Z(x^n, Y^n)}\varphi(w)\prod_{i=1}^{n} p(Y_i|x_i, w), \tag{1.31}$$

$$p(y|x, x^n, Y^n) = \int p(y|x, w)p(w|x^n, Y^n)dw. \tag{1.32}$$

In conditional independent cases, the generalization error is defined by the given x^n, because the expected value over x is not defined,

$$G_n = -\frac{1}{n}\sum_{i=1}^{n}\int dy\ q(y|x_i)\log p(y|x_i, x^n, Y^n),$$

$$T_n = -\frac{1}{n}\sum_{i=1}^{n}\log p(Y_i|x_i, x^n, Y^n).$$

Also the generalization and training errors are defined by

$$G_n - S = \frac{1}{n}\sum_{i=1}^{n}\int q(y|X_i)\log\frac{q(y|x_i)}{p(y|x_i, x^n, Y^n)}dy,$$

$$T_n - S_n = \frac{1}{n}\sum_{i=1}^{n}\log\frac{q(Y_i|x_i)}{p(Y_i|x_i, x^n, Y^n)}.$$

Both the cross validation loss and WAIC can be defined by the same forms as the independent case.

$$C_n = \frac{1}{n}\sum_{i=1}^{n}\log\mathbb{E}_w[1/p(Y_i|x_i, w)],$$

$$W_n = T_n + \frac{1}{n}\sum_{i=1}^{n}\mathbb{V}_w[\log p(Y_i|x_i, w)].$$

However, in this case,

$$\mathbb{E}[C_n] \neq \mathbb{E}[G_{n-1}],$$

whereas, in Section 5.5, we show

$$\mathbb{E}[W_n] = \mathbb{E}[G_n] + o(1/n).$$

Hence, even if x^n is dependent, if C_n is asymptotically equivalent to W_n,

$$\mathbb{E}[C_n] = \mathbb{E}[G_n] + o(1/n)$$

also holds, whereas if otherwise, then the cross validation loss is can not be applied to estimating the generalization loss.

Example 4. (1) In some applications, regression problems of $\{Y_i\}$ for a given fixed set $\{x_i; i = 1, 2, ..., n\}$ are studied, then the cross validation loss cannot be employed.
(2) A time series prediction problem,

$$Z_t = a_1 Z_{t-1} + a_2 Z_{t-2} + a_3 Z_{t-3} + \text{Gaussian noise},$$

can be understood as a regression problem,

$$(Z_{t-1}, Z_{t-2}, Z_{t-3}) = x_t \mapsto Y_t = Z_t.$$

Therefore x_t is dependent, resulting that the cross validation loss cannot be employed, whereas WAIC can be.

1.9   Problems

1. Let $w_0 = (1, 1, 1, ..., 1) \in \mathbb{R}^{10}$ and W be a random variable on $\mathbb{R}^{10}$ which is subject to

$$p(w) = C\{\exp(-||w||^2) + 100 \exp(-10||w - w_0||^2)\},$$

where $||w||^2 = \sum_i w_i^2$ and C is a constant. Show the following:
(1) Let $\hat{w}$ be the maximum point of $p(w)$. Then $\hat{w} \approx w_0$.
(2) $\mathbb{E}[W] \approx 0$.
Hence $\mathbb{E}[W]$ is far from $\hat{w}$. Assume that $p(w)$ is a posterior distribution of some statistical model. Discuss the difference between the maximum likelihood or *a posteriori* estimator and Bayesian estimation.

2. (Fluctuation-dissipation theorem) Let $\beta > 0$ and $H(x)$ be a function from $\mathbb{R}^N$ to $\mathbb{R}$. Assume that a random variable $X \in \mathbb{R}^N$ is subject to a probability density function,

$$p(x|\beta) = \frac{1}{Z(\beta)} \exp(-\beta H(x)),$$

where $Z(\beta)$ is a constant

$$Z(\beta) = \int \exp(-\beta H(x)) dx.$$

Then prove the equation,

$$\frac{\partial \mathbb{E}[H(X)]}{\partial \beta} = -\mathbb{V}[H(X)].$$

It is well known that this equation demonstrates several important laws in physics.

3. (Coin toss) Let $p(x|a)$ be a statistical model of $x \in \{0, 1\}$ which is defined by

$$p(x|a) \;=\; a^x(1 - a)^{1-x},$$

where a is a parameter ($0 \le a \le 1$). See Figure 1.9. Let us study a prior $\varphi(a)$ which is defined by

$$\varphi(a) \;=\; 1,$$

on $0 \le a \le 1$. Let X^n be a set of random variables which are independently subject to $p(x|a_0)$. Also let n_1 and n_2 be

$$n_1 = \sum_{i=1}^{n} X_i, \quad n_2 = n - n_1.$$

Then show the following equations.

(1) The maximum likelihood estimator is $\hat{a} = n_1/n$ and the estimated probability distribution $p(x|\hat{a})$ by the maximum likelihood method is given by

$$p(1|\hat{a}) = \frac{n_1 + \varepsilon}{n + 2\varepsilon}, \quad p(0|\hat{a}) = \frac{n_2 + \varepsilon}{n + 2\varepsilon},$$

where $\varepsilon = 0$.

(2) The Bayesian predictive distribution $p(x|X^n)$ is

$$p(1|X^n) = \frac{n_1 + 1}{n + 2}, \quad p(0|X^n) = \frac{n_2 + 1}{n + 2}.$$

(3) The generalization error of the maximum likelihood method is defined by Kullback-Leibler information from $p(x|a_0)$ to $p(x|\hat{a})$,

$$\begin{aligned} K_{ML} = \quad &-a_0 \log((n_1 + \varepsilon)/(n + 2\varepsilon)) - (1 - a_0) \log((n_2 + \varepsilon)/(n + 2\varepsilon)) \\ &+ a_0 \log a_0 + (1 - a_0) \log(1 - a_0), \end{aligned}$$

where $\varepsilon > 0$ is a sufficiently small positive value. Note that, if $\varepsilon = 0$, $0 < a_0 < 1$, and $n_1 n_2 = 0$, then $K_{ML} = \infty$. That of Bayesian method is defined by Kullback-Leibler distance from $p(x|a_0)$ to $p(x|X^n)$,

$$\begin{aligned} K_{Bayes} = \quad &-a_0 \log((n_1 + 1)/(n + 2)) - (1 - a_0) \log((n_2 + 1)/(n + 2)) \\ &+ a_0 \log a_0 + (1 - a_0) \log(1 - a_0). \end{aligned}$$

(4) By using numerical calculation, the expected values of $\mathbb{E}[K_{ML}]$ and $\mathbb{E}[K_{Bayes}]$ for $n = 20$, $a_0 = 0.05, 0.10, ..., 0.50$, and $\varepsilon = 0.0001$ are shown in Figure 1.9. If $a_0 = 0$, then $\mathbb{E}[K_{ML}] = 0$ and $\mathbb{E}[K_{Bayes}] > 0$. Discuss the difference between the maximum likelihood and Bayesian methods from the veiwpoint of the generalization errors.

4. (Simple normal distribution) Let $p(x|a)$ and $\varphi(a)$ be a statistical model of $x \in \mathbb{R}$ for a given parameter $a \in \mathbb{R}$ and a prior of a respectively,

$$p(x|a) = \frac{1}{(2\pi)^{1/2}} \exp\left(-\frac{1}{2}(x - a)^2\right),$$

$$\varphi(a) = \left(\frac{A}{2\pi}\right)^{1/2} \exp\left(-\frac{A}{2}a^2\right).$$

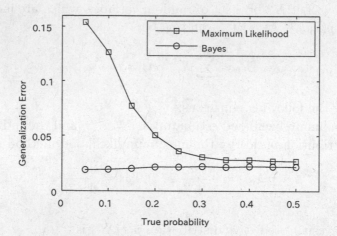

Figure 1.7: Comparison of Bayes with maximum likelihood in coin toss. The averages of the generalization errors by the maximum likelihood and Bayesian methods are compared in a coin toss problem for $n = 20$. The standard deviations of the maximum likelihood method are larger than that of Bayes for every a_0.

Let X^n be a set of independent random variables which are subject to $p(x|0)$. Then prove the following equations.

$$
S = C_0 + \frac{1}{2},
$$

$$
S_n = C_0 + \frac{1}{2n} \sum_{i=1}^{n} x_i^2,
$$

where $C_0 = (1/2) \log(2\pi)$. The losses are

$$
G_n = C_0 + \frac{1}{2} \log(1 + \frac{1}{n_1}) + \frac{n_1}{2(n_1 + 1)} + \frac{n_1}{2(n_1 + 1)}(x^*)^2
$$

$$
C_n = C_0 - \frac{1}{2} \log(1 - \frac{1}{n_1}) + \frac{n_1}{2(n_1 - 1)} \frac{1}{n} \sum_{j=1}^{n} (x_j - x^*)^2
$$

$$
T_n = C_0 + \frac{1}{2} \log(1 + \frac{1}{n_1}) + \frac{n_1}{2(n_1 + 1)} \frac{1}{n} \sum_{j=1}^{n} (x_j - x^*)^2
$$

$$
W_n = C_0 + \frac{1}{2} \log(1 + \frac{1}{n_1}) + \frac{1}{2n_1^2} + \frac{n_1^2 + 2n_1 + 2}{2n_1(n_1 + 1)} \frac{1}{n} \sum_{j=1}^{n} (x_j - x^*)^2,
$$

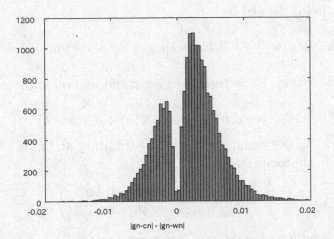

Figure 1.8: Cross validation and WAIC. This figure shows the histogram of $|g_n - c_n| - |g_n - w_n|$. If $|g_n - c_n| - |g_n - w_n| > 0$ then WAIC is a better approximator of the generalization loss than the cross validation.

where $n_1 = n + A$ and

$$x^* = \frac{1}{n_1} \sum_{i=1}^{n} x_i.$$

The free energy is

$$F_n = nC_0 + \frac{1}{2} \log(n_1/A) - \frac{n_1}{2}(x^*)^2 + \frac{1}{2}(\sum_{i=1}^{n} x_i^2).$$

It follows that $C_n - W_n = O_p(1/n^3)$. Let

$$g_n = G_n - S,$$
$$c_n = C_n - S_n,$$
$$w_n = W_n - S_n.$$

Then the histogram of $|g_n - c_n| - |g_n - w_n|$ for $n = 5$ is given by Figure 1.8. In this model, Bayesian observables can be explicitly calculated without numerical approximation, hence it is easy to compare the cross validation loss and WAIC as estimators of the generalization loss.

5. (Normal mixture) The Fisher information matrix $I_{jk}(w)$ of a statistical

model $p(x|w)$ is defined by

$$I_{jk}(w) = \int \partial_j \log p(x|w) \; \partial_k \log p(x|w) \; p(x|w) \; dx$$

where $\partial_j = \partial/\partial w_j$. Show that the Fisher information matrix of a statistical model

$$p(x|a,b) = (1-a)N(x) + aN(x-b) \tag{1.33}$$

where $N(x)$ is the standard normal distribution at $(a_0, b_0) = (0.5, 0.3)$ is numerically approximated by

$$I(a_0, b_0) \approx \begin{pmatrix} 0.0881 & 0.1467 \\ 0.1467 & 0.2552 \end{pmatrix},$$

whose minimum and maximum eigenvalues are

$$0.0029 << 0.3405.$$

Note that the minimum value is far smaller than the maximum value. If a sample X^n is independently subject to $p(x|a_0, b_0)$ and if regular asymptotic theory held, then the asymptotic posterior distribution would be approximated by the normal distribution whose average is (a_0, b_0) and covariance matrix is $(nI(a_0, b_0))^{-1}$. By comparing Figures 1.4, 1.5, and 1.6 with the minimum eigenvalue, discuss the sufficiently large n by which regular asymptotic theory holds. Compare its result with the fact that singular asymptotic theory holds even when n is smaller than 100.

6. (Conditional dependent case) A simple linear regression model of $y \in \mathbb{R}$ for given $x \in \mathbb{R}$ and paramater $a \in \mathbb{R}$ is defined by

$$p(y|x,a) = \frac{1}{\sqrt{2\pi}} \exp\left(-\frac{1}{2}(y-ax)^2\right).$$

Assume that $p(y|x, a_0)$ is a true conditional probability density of $y \in \mathbb{R}$. Let $\{x_i; i = 1, 2, ..., n\}$ and $\{\xi_i; i = 1, 2, ..., N\}$ be sets of fixed input data used in estimation (training) and trial (test), respectively. The set of conditional independent data is $\{(x_i, Y_i); i = 1, 2, ..., n\}$. For simple calculation, we employ an improper prior $\varphi(a) = 1$ on $\mathbb{R}$. Then the posterior and predictive distributions are respectively given by

$$p(a|x^n, Y^n) = \frac{1}{Z} \prod_{i=1}^{n} p(Y_i|x_i, a),$$
$$p^*(y|x) = \mathbb{E}_a[p(y|x,a)],$$

where Z is a constant. The generalization error is defined by using the test set,

$$G_n = -\frac{1}{N} \sum_{j=1}^{N} \int p(y|\xi_j, a_0) \log p^*(y|\xi_j) dy,$$

which is a random variable because the predictive distribution is a function of Y^n. The leave-one-out cross validation loss and the widely applicable information criterion for a set are respectively defined by using the training set,

$$C_n = \frac{1}{n} \sum_{i=1}^{n} \log \mathbb{E}_a \left[\frac{1}{p(Y_i|x_i, a)} \right],$$

$$W_n = -\frac{1}{n} \sum_{i=1}^{n} \log p^*(Y_i|x_i) + \frac{1}{n} \sum_{i=1}^{n} \mathbb{V}_a[\log p(Y_i|x_i, a)].$$

The conditional true entropy is

$$S = -\frac{1}{n} \sum_{i=1}^{n} \int p(y|x_i, a_0) \log p(y|x_i, a_0) dy = \frac{1}{2} \log(2\pi) + \frac{1}{2}.$$

Then show the following equations.

$$\mathbb{E}[G_n] - S = \frac{1}{2N} \sum_{j=1}^{N} \log\left(1 + \frac{\xi_j^2}{\sum_{i=1}^{n} x_i^2}\right),$$

$$\mathbb{E}[C_n] - S = -\frac{1}{2n} \sum_{j=1}^{n} \log\left(1 - \frac{x_j^2}{\sum_{i=1}^{n} x_i^2}\right),$$

$$\mathbb{E}[W_n] - S = \frac{1}{2n} \sum_{j=1}^{n} \log\left(1 + \frac{x_j^2}{\sum_{i=1}^{n} x_i^2}\right) + \frac{1}{2n} \sum_{j=1}^{n} \frac{1 - s_j}{1 + s_j} s_j^2,$$

where

$$s_j = \frac{x_j^2}{\sum_{i=1}^{n} x_i^2}.$$

Prove that, if $n = N$ and $\xi_i = x_i$ for all i,

$$\mathbb{E}[G_n] \leq \mathbb{E}[W_n] \leq \mathbb{E}[C_n],$$

where equalities hold if and only if $x_i = 0$ for all i.

Chapter 2

Statistical Models

In this chapter, we introduce several concrete examples of statistical models. The main purpose of this book is to establish the mathematically universal theory in Bayesian statistics which holds even for nonregular statistical models. However, before studying the general formulas, concrete examples are prepared for understanding them. We introduce

(1) Normal distribution
(2) Multinomial distribution
(3) Linear regression
(4) Neural network
(5) Finite normal mixture
(6) Nonparametric mixture

and then examine the behaviors of the free energy or the minus log marginal likelihood, and the generalization, training, cross validation losses, and WAIC. The statistical models (1), (2), and (3) are regular, whereas (4), (5), and (6) are nonregular. If a reader has software for numerical calculation, then it will be easy to realize them.

2.1   Normal Distribution

Firstly, we study a normal distribution. Let us use a probability density function of $x \in \mathbb{R}$ for a given parameter $w = (m, s) \in \mathbb{R}^2$, $(s > 0)$.

$$p(x|m, s) \;=\; \sqrt{\frac{s}{2\pi}} \exp\Big(-\frac{s}{2}(x - m)^2\Big). \tag{2.1}$$

The probability distribution represented by this density function is denoted by $\mathcal{N}(m, 1/s)$, where m is the average and $1/s$ is the variance. It can be rewritten as

$$p(x|m,s) \;=\; \frac{1}{\sqrt{2\pi}} \exp\Big(-\frac{s}{2}x^2 + msx - \frac{m^2 s}{2} + \frac{1}{2}\log s\Big).$$

The conjugate prior $\varphi(m,s|\phi_1,\phi_2,\phi_3)$ of the normal distribution is the same function of a parameter as the statistical model, by replacing $(x^2,x,1)$ by a hyperparameter $\phi = (\phi_1,\phi_2,\phi_3)$,

$$
\begin{aligned}
\varphi(w|\phi) \;&=\; \varphi(m,s|\phi_1,\phi_2,\phi_3) \\
&=\; \frac{1}{Y(\phi)} \exp\Big(-\frac{s}{2}\phi_1 + ms\phi_2 - \frac{1}{2}(m^2 s - \log s)\phi_3\Big), \quad (2.2)
\end{aligned}
$$

where $w = (m,s)$. It follows that

$$\varphi(w|\phi) \;=\; \frac{1}{Y(\phi)} \exp\Big(-\frac{s}{2}\big(\phi_1 - \frac{\phi_2^2}{\phi_3}\big)\Big) \Big(s^{1/2}\exp(-\frac{s}{2}(m - \frac{\phi_2}{\phi_3})^2)\Big)^{\phi_3}.$$

The constant $Y(\phi)$ is determined by the condition $\int \varphi(w|\phi)dw = 1$. By using integral formulas,

$$\int_{-\infty}^{\infty} \exp(-ax^2)dx \;=\; \sqrt{\pi/a}, \qquad\qquad (2.3)$$

$$\int_{0}^{\infty} x^{a-1}\exp(-x/b)dx \;=\; b^a\,\Gamma(a), \qquad\qquad (2.4)$$

where $\Gamma(a)$ is the gamma function, the function $Y(\phi)$ is given by

$$Y(\phi) = \frac{2\sqrt{\pi}(2\phi_3)^{\phi_3/2}}{(\phi_1\phi_3 - \phi_2^2)^{(\phi_3+1)/2}} \Gamma\Big(\frac{\phi_3+1}{2}\Big). \qquad\qquad (2.5)$$

To ensure $0 < Y(\phi) < \infty$, the hyperparameter should satisfy $\phi_3 > 0$ and $\phi_1\phi_3 - \phi_2^2 > 0$. Since the conjugate prior has the same form of the parameter as the statistical model, the posterior simultaneous density of (w, X^n) is given by

$$
\begin{aligned}
\Omega(w, X^n) \;&\equiv\; \varphi(w)\prod_{i=1}^{n} p(X_i|w) \\
&=\; \frac{1}{Y(\phi)(2\pi)^{n/2}} \exp\Big(-\frac{s}{2}\hat{\phi}_1 + ms\hat{\phi}_2 - \frac{1}{2}(m^2 s - \log s)\hat{\phi}_3\Big),
\end{aligned}
$$

where

$$\hat{\phi}_1 = \sum_{i=1}^{n} X_i^2 + \phi_1, \tag{2.6}$$

$$\hat{\phi}_2 = \sum_{i=1}^{n} X_i + \phi_2, \tag{2.7}$$

$$\hat{\phi}_3 = n + \phi_3. \tag{2.8}$$

The partition function is given by

$$Z(X^n) = \int \Omega(w, X^n) dw = \frac{Y(\hat{\phi})}{Y(\phi)(2\pi)^{n/2}},$$

and the posterior distribution, which is equal to $\Omega(w, X^n)/Z(X^n)$, is given by

$$p(w|X^n) = \varphi(m, s|\hat{\phi}_1, \hat{\phi}_2, \hat{\phi}_3).$$

Hence the minus log marginal likelihood or the free energy $F = -\log Z(X^n)$ is

$$F_n = \frac{n}{2} \log(2\pi) + \log Y(\phi) - \log Y(\hat{\phi}). \tag{2.9}$$

See Figure 2.1. The predictive density is also given by

$$\begin{aligned}
\mathbb{E}_w[p(x|w)] &= \int p(x|w)\varphi(w|\hat{\phi})dw \\
&= \frac{1}{\sqrt{2\pi}} \frac{Y(\hat{\phi}_1 + x^2, \hat{\phi}_2 + x, \hat{\phi}_3 + 1)}{Y(\hat{\phi}_1, \hat{\phi}_2, \hat{\phi}_3)} \\
&\propto \frac{1}{[(x - \hat{\phi}_2/\hat{\phi}_3)^2 + C_1]^{(\hat{\phi}_3+1)/2}},
\end{aligned} \tag{2.10}$$

where $C_1 = (\hat{\phi}_1\hat{\phi}_3 - \hat{\phi}_2^2)(\hat{\phi}_3 + 1)/\hat{\phi}_3^2$. The predictive density is different from any normal distribution. However, when $n \to \infty$, it converges to a normal distribution. See Figure 2.2. The training loss is

$$\begin{aligned}
T_n &= -\frac{1}{n} \sum_{i=1}^{n} \log \mathbb{E}_w[p(X_i|w)] \\
&= \frac{1}{2} \log(2\pi) + \log Y(\hat{\phi}_1, \hat{\phi}_2, \hat{\phi}_3) \\
&\quad - \frac{1}{n} \sum_{i=1}^{n} \log Y(\hat{\phi}_1 + X_i^2, \hat{\phi}_2 + X_i, \hat{\phi}_3 + 1).
\end{aligned} \tag{2.11}$$

Since

$$\mathbb{E}_w[1/p(x|w)] = \sqrt{2\pi} \, \frac{Y(\hat{\phi}_1 - x^2, \hat{\phi}_2 - x, \hat{\phi}_3 - 1)}{Y(\hat{\phi}_1, \hat{\phi}_2, \hat{\phi}_3)},$$

the cross validation loss is equal to

$$
\begin{aligned}
C_n &= \frac{1}{n} \sum_{i=1}^{n} \log \mathbb{E}_w[1/p(X_i|w)] \\
&= \frac{1}{2} \log(2\pi) - \log Y(\hat{\phi}_1, \hat{\phi}_2, \hat{\phi}_3) \\
&\quad + \frac{1}{n} \sum_{i=1}^{n} \log Y(\hat{\phi}_1 - X_i^2, \hat{\phi}_2 - X_i, \hat{\phi}_3 - 1).
\end{aligned}
\tag{2.12}
$$

Let $f(\alpha, x)$ be a function

$$
\begin{aligned}
f(\alpha, x) &= \log \int p(x|w)^{\alpha} \varphi(w|\hat{\phi}) dw \\
&= \log \Big[\frac{1}{(2\pi)^{\alpha/2}} \, \frac{Y(\hat{\phi}_1 + \alpha x^2, \hat{\phi}_2 + \alpha x, \hat{\phi}_3 + \alpha)}{Y(\hat{\phi}_1, \hat{\phi}_2, \hat{\phi}_3)} \Big].
\end{aligned}
$$

Since the posterior distribution is equal to $\varphi(w|\hat{\phi})$,

$$
\begin{aligned}
\frac{\partial^2 f}{\partial \alpha^2}(0, x) &= \mathbb{V}_w[\log p(x|w)] \\
&= \frac{\partial^2}{\partial \alpha^2} \Big[\frac{1}{2}(\hat{\phi}_3 + \alpha) \log(2(\hat{\phi}_3 + \alpha)) \\
&\quad - \frac{1}{2}(\hat{\phi}_3 + \alpha + 1) \log\{(\hat{\phi}_1 + \alpha x^2)(\hat{\phi}_3 + \alpha) - (\hat{\phi}_2 + \alpha x)^2\} \\
&\quad + \log \Gamma((\hat{\phi}_3 + \alpha + 1)/2) \Big] \Big|_{\alpha=0} \\
&= \frac{1}{4} \psi'\Big(\frac{\hat{\phi}_3 + 1}{2}\Big) + \frac{1}{2\hat{\phi}_3} - u(x) + \frac{(\hat{\phi}_3 + 1)u(x)^2}{2},
\end{aligned}
$$

where $\psi'(x) = (\log \Gamma(x))''$ is the trigamma function and

$$u(x) = \frac{\hat{\phi}_3 x^2 - 2\hat{\phi}_2 x + \hat{\phi}_1}{\hat{\phi}_1 \hat{\phi}_3 - (\hat{\phi}_2)^2}.$$

Therefore WAIC is given by

$$W_n = T_n + \frac{1}{n} \sum_{i=1}^{n} \frac{\partial^2 f}{\partial \alpha^2}(0, X_i).$$

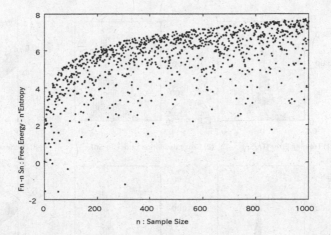

Figure 2.1: Free energy for $n = 1, 2, ..., 1000$. The free energy or the minus log marginal likelihood of a normal distribution $F_n - nS_n$ is shown for $n = 1, 2, ..., 1000$. Its asymptotic behavior is given by $\log n$.

By using the function $Y(\phi)$ in eq.(2.5), we can calculate F_n, T_n, and C_n using X^n, based on eqs.(2.9), (2.11), and (2.12). Let us assume that a true distribution is $q(x) = p(x|m_0, s_0)$. Then the entropy and empirical entropy of the true distribution are respectively given by

$$
\begin{aligned}
S &= -\int q(x)\{\frac{1}{2}\log(\frac{s_0}{2\pi}) - \frac{s_0}{2}(x - m_0)^2\}dx \\
&= -\frac{1}{2}\log(\frac{s_0}{2\pi}) + \frac{1}{2}, \\
S_n &= -\frac{1}{2}\log(\frac{s_0}{2\pi}) + \frac{s_0}{2n}\sum_{i=1}^{n}(X_i - m_0)^2.
\end{aligned}
$$

The training and cross validation errors are respectively given by $T_n - S_n$ and $C_n - T_n$. The generalization loss is given by

$$
\begin{aligned}
G_n &= S + \int q(x)\log\frac{q(x)}{\mathbb{E}_w[p(x|w)]}dx \\
&= \frac{1}{2}\log(2\pi) + \log Y(\hat{\phi}_1, \hat{\phi}_2, \hat{\phi}_3) \\
&\quad - \int q(x)\log Y(\hat{\phi}_1 + x^2, \hat{\phi}_2 + x, \hat{\phi}_3 + 1)dx.
\end{aligned}
$$

Unfortunately, the integration over $q(x)$ cannot be done analytically.

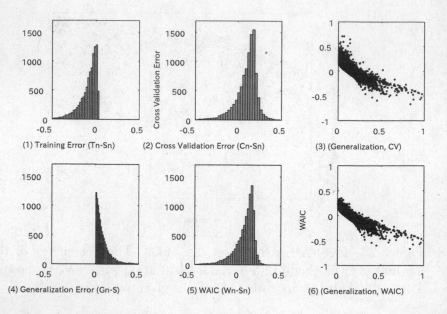

Figure 2.2: Bayesian observables in normal distribution are shown in the case $n = 10$. Histograms of (1) training error $T_n - S_n$, (2) cross validation error $C_n - S_n$, (4) generalization error $G_n - S$, and (5) WAIC error $W_n - S_n$. Distributions of (3) $(G_n - S, C_n - S_n)$, and (6) $(G_n - S, W_n - S_n)$. Note that WAIC error has smaller variance than cross validation error.

Example 5. A numerical experiment was conducted by setting $(m_0, s_0) = (1, 1)$ and $(\phi_1, \phi_2, \phi_3) = (0.5, 0, 0.5)$. In Figure 2.1 the horizontal and vertical axes show the sample size n and the free energy or the minus log marginal likelihood minus empirical entropy $F_n - nS_n$ for $n = 1, 2, ..., 1000$, respectively. In the following sections, we will show that $F_n - nS_n = (d/2) \log n + O_p(1)$ (d is the dimension of the parameter space), which is consistent with the figure. In Figure 2.2, experimental results of 10000 independent trials for $n = 10$ are shown.

(1) Histogram of the training error, $T_n - S_n$

(2) Cross validation error, $C_n - S_n$

(3) Generalization error and the cross validation error

(4) Histogram of generalization error, $G - S$

(5) Histogram of WAIC error, $W_n - S_n$

(6) Generalization error and WAIC

The averages and standard deviations of $G_n - S$, $C_n - S_n$, and $W_n - S_n$ are numerically approximated by

$$\text{Average:} \quad 0.0901, \ 0.1017, \ 0.0860,$$
$$\text{Standard deviation:} \quad 0.0978, \ 0.1211, \ 0.1136.$$

The variance of the cross validation error is larger than WAIC error. Moreover,

$$\mathbb{E}|(G_n - S) - (C_n - S_n)| \ = \ 0.153,$$
$$\mathbb{E}|(G_n - S) - (W_n - S_n)| \ = \ 0.146.$$

Therefore, WAIC is the better estimator of the generalization error than the cross validation error in this case. The inequality

$$\mathbb{E}|(G_n - S) - (C_n - S_n)| > \mathbb{E}|(G_n - S) - (W_n - S_n)|$$

holds for $n = 1, 2, 3, 4, 5$. In the following chapters, we show the higher order asymptotic equivalence of the cross validation and WAIC as $n \to \infty$. For the finite and smaller n, WAIC is the better approximator of the generalization loss than the cross validation loss in many statistical models.

2.2 Multinomial Distribution

A multinomial distribution is examined. Let N be a positive integer. An N dimensional variable

$$x = (x^{(1)}, x^{(2)}, ..., x^{(N)})$$

is said to be competitive if only one element is equal to one, $x^{(j)} = 1$, and others are equal to zero, $x^{(k)} = 0 \ (k \neq j)$. The set of N dimensional competitive variables is denoted by

$$\mathcal{C}_N \equiv \{x = (x^{(1)}, x^{(2)}, \cdots, x^{(N)}); \ x \text{ is competitive}\}.$$

By the definition, the number of elements of the set $\mathcal{C}_N$ is equal to N. This set is used in classification problems where N is the number of categories. In a coin toss problem, $N = 2$, whereas in a dice throw problem, $N = 6$. The N dimensional multinomial distribution of one trial $x \in \mathcal{C}_N$ for a given parameter w is defined by

$$p(x|w) = \prod_{j=1}^{N} (w_j)^{x_j}, \tag{2.13}$$

where we determine $0^0 = 1$ and a set of all parameters is

$$W = \{w = (w_1, w_2, \cdots, w_N) \, ; \, \sum_{j=1}^{N} w_j = 1, \, w_i \geq 0\}.$$

If a random variable $X = (X^{(1,)} X^{(2)}, ..., X^{(N)})$ is subject to $p(x|w)$, then for an arbitrary j, the probability $X^{(j)} = 1$ is equal to

$$\text{Prob}(X^{(j)} = 1) = w_j.$$

The Dirichlet distribution on W is often employed as a prior,

$$\varphi(w|a) = \frac{1}{C(a)} \prod_{j=1}^{N} (w_j)^{a_j - 1} \, \delta\Big(\sum_{j=1}^{N} w_j - 1\Big), \qquad (2.14)$$

where a hyperparameter a is an N dimensional vector,

$$a = (a_1, a_2, ..., a_N),$$

which satisfies $a_j > 0 \; (j = 1, 2, ..., N)$ and

$$C(a) \;=\; \Big(\prod_{j=1}^{N} \int_0^1 dw_j \, (w_j)^{a_j - 1}\Big) \delta\Big(\sum_{j=1}^{N} w_j - 1\Big).$$

Lemma 1. *The normalizing term of Dirichlet distribution is*

$$C(a) \;=\; \frac{\displaystyle\prod_{j=1}^{N} \Gamma(a_j)}{\Gamma\Big(\displaystyle\sum_{j=1}^{N} a_j\Big)}. \qquad (2.15)$$

Proof. Let $\lambda > 0$ and

$$f(\lambda) \;=\; \Big(\prod_{j=1}^{N} \int_0^\infty dw_j \, (w_j)^{a_j - 1}\Big) \delta\Big(\sum_{j=1}^{N} a_j - \lambda\Big).$$

Then $f(1) = C(a)$. The Laplace transform of $f(\lambda)$ is

$$\int_0^\infty f(\lambda)e^{-\beta\lambda}d\lambda = \prod_{j=1}^N \int_0^\infty (w_j)^{a_j-1} \exp(-\beta w_j)dw_j$$

$$= \prod_{j=1}^N \frac{1}{(\beta)^{a_j}} \int_0^\infty (w_j)^{a_j-1} \exp(-w_j)dw_j$$

$$= \frac{1}{(\beta)^{\sum_j a_j}} \prod_{j=1}^N \Gamma(a_j)$$

$$= \frac{\prod_j \Gamma(a_j)}{\Gamma(\sum_j a_j)} \int_0^\infty \lambda^{\sum_j a_j-1} e^{-\beta\lambda}d\lambda.$$

Therefore, by using the inverse Laplace transform of this equation, $f(\lambda)$ is obtained,

$$f(\lambda) = \frac{\prod_j \Gamma(a_j)}{\Gamma(\sum_j a_j)} \lambda^{\sum_j a_j-1}.$$

Then $f(1)$ gives eq.(2.15). $\qquad\qquad\qquad\qquad\qquad\qquad\qquad\square$

Let us make the posterior distribution for a multinomial distribution and Dirichlet prior. Let X^n be a set of n random variables on $\mathcal{C}_N$,

$$X^n = \{X_i = (X_i^{(1)}, ..., X_i^{(N)}) \; ; \; i = 1, 2, ..., n\}.$$

A random variable n_j is defined by

$$n_j = \sum_{i=1}^n X_i^{(j)},$$

which is the number of sample points classified into the jth category. Then $\sum_j n_j = n$ and the posterior distribution is

$$p(w|X^n) = \frac{1}{Z(X^n)} \frac{1}{C(a)} \prod_{j=1}^N (w_j)^{a_j-1} \prod_{i=1}^n (w_j)^{X_i^{(j)}} \delta\left(\sum_{j=1}^N w_j - 1\right)$$

$$= \frac{1}{Z(X^n)} \frac{1}{C(a)} \prod_{j=1}^N (w_j)^{n_j+a_j-1} \delta\left(\sum_{j=1}^N w_j - 1\right).$$

By using a notation $\overline{n} = (n_1, n_2, ..., n_N)$, eq. (2.15), and $\int p(w|X^n)dw = 1$, the marginal likelihood or the partition function satisfies

$$Z(X^n)C(a) = C(\overline{n} + a).$$

Therefore

$$Z(X^n) = \frac{\prod_{j=1}^{N} \Gamma(n_j + a_j)}{\Gamma(n + \sum_{j=1}^{N} a_j)} \cdot \frac{\Gamma(\sum_{j=1}^{N} a_j)}{\prod_{j=1}^{N} \Gamma(a_j)}.$$

Thus the minus log marginal likelihood or the free energy is given by

$$
\begin{aligned}
F_n &= \log \Gamma(n + \sum_{j=1}^{N} a_j) - \sum_{j=1}^{N} \log \Gamma(n_j + a_j) \\
&\quad - \log \Gamma(\sum_{j=1}^{N} a_j) + \sum_{j=1}^{N} \log \Gamma(a_j).
\end{aligned}
\tag{2.16}
$$

The predictive distribution $p(x|X^n)$ of $x = (x^{(1)}, x^{(2)}, ..., x^{(N)})$ is given by

$$
\begin{aligned}
p(x|X^n) &= E_w[p(x|w)] = \int p(x|w)p(w|X^n)dw \\
&= \frac{Z(X^{n+1})}{Z(X^n)} \\
&= \frac{\prod_{j=1}^{N} \Gamma(x^{(j)} + n_j + a_j)}{\Gamma(n + 1 + \sum_{j=1}^{N} a_j)} \cdot \frac{\Gamma(n + \sum_{j=1}^{N} a_j)}{\prod_{j=1}^{N} \Gamma(n_j + a_j)},
\end{aligned}
$$

where we used a notation $X^{n+1} = (X^n, x)$. By using $\Gamma(x + 1) = x\Gamma(x)$ for an arbitrary $x > 0$, it follows that

$$p(x|X^n) = \frac{\prod_{j=1}^{N} (n_j + a_j)^{x^{(j)}}}{n + \sum_{j=1}^{N} a_j}.$$

Thus the training loss is

$$
\begin{aligned}
T_n &= -\frac{1}{n} \sum_{i=1}^{n} \log p(X_i|X^n) \\
&= \log(n + \sum_{j=1}^{N} a_j) - \frac{1}{n} \sum_{i=1}^{n} \sum_{j=1}^{N} X_{ij} \log(n_j + a_j) \\
&= \log(n + \sum_{j=1}^{N} a_j) - \sum_{j=1}^{N} \frac{n_j}{n} \log(n_j + a_j).
\end{aligned}
$$

A part of the cross validation loss is

$$\mathbb{E}_w[1/p(X_i|w)] = \frac{Z(X^n \setminus X_i)}{Z(X^n)}$$

$$= \frac{\prod_{j=1}^{N} \Gamma(-X_i^{(j)} + n_j + a_j)}{\Gamma(n - 1 + \sum_{j=1}^{N} a_j)} \cdot \frac{\Gamma(n + \sum_{j=1}^{N} a_j)}{\prod_{j=1}^{N} \Gamma(n_j + a_j)}$$

$$= \frac{n - 1 + \sum_{j=1}^{N} a_j}{\prod_{j=1}^{N}(n_j + a_j - 1)^{X_i^{(j)}}}.$$

Hence the cross validation loss is given by

$$C_n = \frac{1}{n}\sum_{i=1}^{n} \log \mathbb{E}_w[1/p(X_i|X^n)]$$

$$= \log(n - 1 + \sum_{j=1}^{N} a_j) - \frac{1}{n}\sum_{i=1}^{n}\sum_{j=1}^{N} X_i^{(j)} \log(n_j + a_j - 1)$$

$$= \log(n - 1 + \sum_{j=1}^{N} a_j) - \sum_{j=1}^{N} \frac{n_j}{n} \log(n_j - 1 + a_j). \qquad (2.17)$$

The above equations for F_n, T_n, C_n, and W_n can be applicable in any sample X^n. However, if we adopt the hyperparameter $a = (a_1, a_2, ..., a_N)$ as $0 < a_j \leq 1$, and if at least one of n_j $(j = 1, 2, ..., N)$ is equal to zero, then the cross validation loss diverges. In practical applications, we had better remark that, if N is large or n is small, the data contains $n_j = 0$. Then by using

$$\int p(x|w)^{\alpha} p(w|X^n) dw = \frac{\Gamma(n + \sum_j a_j) \prod_j \Gamma(\alpha x^{(j)} + n_j + a_j)}{\Gamma(n + \alpha + \sum_j a_j) \prod_j \Gamma(n_j + a_j)}, \qquad (2.18)$$

where $\sum_j$ and $\prod_j$ represent the sum and the product for $j = 1, 2, 3, ...N$ respectively, we can derive

$$\mathbb{V}_w[\log p(x|w)] = \frac{\partial^2}{\partial \alpha^2}\Big[\log \int p(x|w)^{\alpha} p(w|X^n) dw\Big]_{\alpha=0}$$

$$= \sum_{j=1}^{N} (x^{(j)})^2 \psi'(n_j + a_j) - \psi'(n + \sum_{j=1}^{N} a_j)$$

where $\psi'(x) = (\log \Gamma(x))''$ is the trigamma function. Therefore WAIC is given by

$$W_n = T_n + V_n,$$

where

$$V_n = \sum_{j=1}^{N} \frac{n_j}{n} \psi'(n_j + a_j) - \psi'(n + \sum_{j=1}^{N} a_j).$$

Assume that $p(x|w_0)$ is the true distribution, where

$$w_0 = (w_{01}, w_{02}, ..., w_{0N}).$$

The generalization loss is explicitly given by

$$
\begin{aligned}
G_n &= -\sum_{x} p(x|w_0) \log p(x|X^n) \\
&= -\sum_{x} \log\Big(\frac{\prod_{j=1}^{N}(n_j + a_j)^{x^{(j)}}}{n + \sum_{j=1}^{N} a_j}\Big) \prod_{j=1}^{N} (w_{0j})^{x^{(j)}} \\
&= \log(n + \sum_{j=1}^{N} a_j) - \sum_{x}\{\sum_{j=1}^{n} x^{(j)} \log(n_j + a_j)\} \prod_{j=1}^{N}(w_{0j})^{x^{(j)}} \\
&= \log(n + \sum_{j=1}^{N} a_j) - \sum_{j=1}^{N} w_{0j} \log(n_j + a_j).
\end{aligned}
\tag{2.19}
$$

The entropy and the empirical entropy of the true distribution are respectively given by

$$
\begin{aligned}
S &= -\sum_{x} p(x|w_0) \log p(x|w_0) \\
&= -\sum_{x} \prod_{j=1}^{n} (w_{0j})^{x^{(j)}} \log(\prod_{j=1}^{n}(w_{0j})^{x^{(j)}}) \\
&= -\sum_{j=1}^{N} w_{0j} \log w_{0j},
\end{aligned}
\tag{2.20}
$$

$$
\begin{aligned}
S_n &= -\frac{1}{n} \sum_{i=1}^{n} \log p(X_i|w_0) \\
&= -\sum_{j=1}^{N} \frac{n_j}{n} \log w_{0j}.
\end{aligned}
\tag{2.21}
$$

Hence the generalization error can be calculated.

Example 6. An experiment was conducted for the case $N = 5$, $w_0 = (0.1, 0.15, 0.2, 0.25, 0.3)$, and $a = (1.1, 1.1, 1.1, 1.1, 1.1)$. Figure 2.3 shows

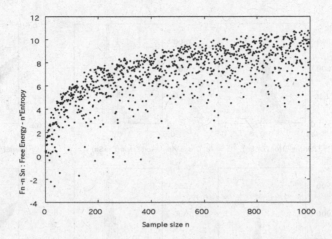

Figure 2.3: $F_n - nS_n$ for $n = 1, 2, ..., 1000$. The free energy or the minus log marginal likelihood of the multinomial distribution is shown. The asymptotic behavior of the random variable $F_n - nS_n$ is in proportion to $\log n$.

experimental results of $F_n - nS_n$ $n = 1, 2, 3, .., 1000$. Figure 2.4 shows experimental results for $n = 20$.

(1) Histogram of the training error, $T_n - S_n$
(2) Cross validation error, $C_n - S_n$
(3) Generalization error and the cross validation error
(4) Histogram of the generalization error, $G - S$
(5) Histogram of WAIC error, $W_n - S_n$
(6) Generalization error and WAIC

The averages and standard deviations of $G_n - S$, $C_n - S_n$, and $W_n - S_n$ are numerically approximated by 10000 independent trials.

$$\text{Average:} \quad 0.0684, \quad 0.0700, \quad 0.0672,$$
$$\text{Standard deviation:} \quad 0.0502, \quad 0.0749, \quad 0.0749.$$

The variance of the cross validation error is almost same as the WAIC error. Moreover,

$$\mathbb{E}|(G_n - S) - (C_n - S_n)| = 0.0948,$$
$$\mathbb{E}|(G_n - S) - (W_n - S_n)| = 0.0943.$$

Therefore, in this case the cross validation error is almost same as the WAIC error.

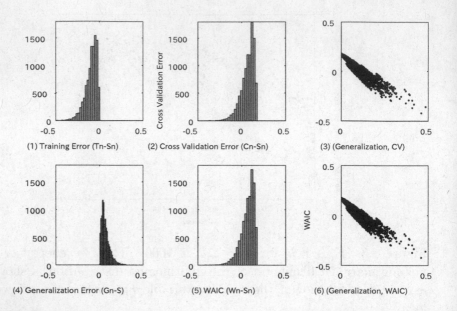

Figure 2.4: Bayesian observables of multinomial distribution for $n = 20$. Bayesian observables of multinomial distribution are shown with $n = 20$. Histograms of (1) training error $T_n - S_n$, (2) cross validation error $C_n - S_n$, (4) generalization error $G_n - S$, and (5) WAIC error $W_n - S_n$. Distributions of (3) $(G_n - S, C_n - S_n)$, and (6) $(G_n - S, W_n - S_n)$.

2.3   Linear Regression

In the foregoing two models, the posterior averaging could be done analytically. In this section, we study a linear regression model, where the posterior averaging is numerically approximated by random sampling. Let us study a statistical model and a prior which are defined for $x, y, a \in \mathbb{R}^1$, $s > 0$ by

$$p(y|x, a, s) = \sqrt{\frac{s}{2\pi}} \exp(-\frac{s}{2}(y - ax)^2),$$

$$\varphi(a, s|r) = \frac{1}{Y(0, 1, 1)} s^r \exp(-\frac{s}{2}(a^2 + 1)),$$

where r is a hyperparameter. This prior is proper if and only if $r > -1/2$. The normalizing constant $Y(\ell, \mu, \rho)$ is given by

$$Y(\ell, \mu, \rho) = \int_{-\infty}^{\infty} da \int_0^{\infty} ds \, s^{r+\ell/2} \exp(-\frac{s}{2}(\mu a^2 + \rho)).$$

Lemma 2. *The normalizing constant* $Y(\ell, \mu, \rho)$ *is equal to*

$$Y(\ell, \mu, \rho) = (2\pi/\mu)^{1/2}(2/\rho)^{r+(\ell+1)/2}\Gamma(r + (\ell+1)/2).$$

Proof. By the definition,

$$
\begin{aligned}
Y(\ell, \mu, \rho) &= \int_{-\infty}^{\infty} da \int_0^{\infty} ds \, s^{r+\ell/2} \exp(-\frac{s}{2}(\mu a^2 + \rho)) \\
&= \int_0^{\infty} ds \, s^{r+\ell/2}(2\pi/s\mu)^{1/2} \exp(-\frac{s\rho}{2}) \\
&= (2\pi/\mu)^{1/2} \int_0^{\infty} ds \, s^{r+\ell/2-1/2} \exp(-\frac{s\rho}{2}) \\
&= (2\pi/\mu)^{1/2}(2/\rho)^{r+\ell/2+1/2} \int_0^{\infty} ds \, s^{r+\ell/2-1/2} \exp(-s) \\
&= (2\pi/\mu)^{1/2}(2/\rho)^{(1+2r+\ell)/2}\Gamma((\ell + 2r + 1)/2),
\end{aligned}
$$

which completes the lemma. □

Let $(X^n, Y^n) = \{(X_i, Y_i); i = 1, 2, ..., n\}$ be a sample which is independently taken from a true probability density $q(x, y)$. The simultaneous distribution $\Omega(a, s, Y^n | X^n)$ for given X^n using the statistical model is given by

$$
\begin{aligned}
\Omega(a, s, Y^n | X^n) &\equiv \varphi(a, s | r) \prod_{i=1}^{n} p(Y_i | X_i, a, s) \\
&= \frac{s^{n/2+r}}{Y(0, 1, 1)(2\pi)^{n/2}} \exp(-\frac{s}{2}(\sum_{i=1}^{n}(Y_i - aX_i)^2 + 1 + a^2)) \\
&= \frac{s^{n/2+r}}{Y(0, 1, 1)(2\pi)^{n/2}} \exp[-\frac{s}{2}\{A(a - B)^2 + C\}],
\end{aligned}
$$

where A, B, and C are constant functions of the parameter,

$$A = \sum_i X_i^2 + 1,$$

$$B = \frac{\sum_i X_i Y_i}{\sum_i X_i^2 + 1},$$

$$C = \{-\frac{(\sum_i X_i Y_i)^2}{\sum_i X_i^2 + 1} + \sum_i Y_i^2 + 1\}.$$

Note that, even if a prior is improper, the posterior distribution is well-defined if $(n-1)/2 + r > -1$. As a result, the posterior density is

$$p(a, s|X^n, Y^n) = \frac{1}{Z(X^n, Y^n)} \Omega(a, s, Y^n|X^n),$$

where the partition function or the marginal likelihood is

$$Z(X^n, Y^n) = \int \Omega(a, s, Y^n|X^n) da ds = \frac{Y(n, A, C)}{(2\pi)^{n/2} Y(0, 1, 1)}.$$

The free energy or the minus log marginal likelihood is

$$F_n = \frac{n}{2} \log(2\pi) + \log Y(0, 1, 1) - \log Y(n, A, C).$$

Then the posterior density can be decomposed as

$$p(a, s|X^n, Y^n) = p(a|s, X^n, Y^n) \, p(s|X^n), \qquad (2.22)$$

$$p(a|s, X^n, Y^n) \propto \exp(-\frac{sA}{2}(a - B)^2), \qquad (2.23)$$

$$p(s|X^n) \propto s^{(n-1)/2+r} \exp(-\frac{Cs}{2}). \qquad (2.24)$$

Thus a set of parameters $\{(a_t, s_t); t = 1, 2, ..., T\}$ which is independently subject to the posterior density can be obtained by the following procedure. Firstly each variable in $\{s_t\}$ is independently taken from $p(s|X^n)$, then each variable in $\{a_t\}$ is independently taken from $p(a|s, X^n, Y^n)$. Then the posterior average of a function $f(a, s)$ can be numerically approximated by

$$\mathbb{E}_{(a,s)}[f(a, s)] \approx \frac{1}{T} \sum_{t=1}^{T} f(a_t, s_t). \qquad (2.25)$$

The generalization, training, and cross validation losses and WAIC are approximated respectively by

$$G_n \approx - \int q(y|x) q(x) \log\Big(\frac{1}{T} \sum_{t=1}^{T} p(y|x, a_t, s_t)\Big) dx dy,$$

$$T_n \approx - \frac{1}{n} \sum_{i=1}^{n} \log\Big(\frac{1}{T} \sum_{t=1}^{T} p(Y_i|X_i, a_t, s_t)\Big),$$

$$C_n \approx \frac{1}{n} \sum_{i=1}^{n} \log\Big(\frac{1}{T} \sum_{t=1}^{T} 1/p(Y_i|X_i, a_t, s_t)\Big),$$

$$W_n \approx T_n + V_n,$$

where V_n is the sum of the posterior variance,

$$V_n = \frac{1}{n} \sum_{i=1}^{n} \left\{ \frac{1}{T} \sum_{t=1}^{T} (\log p(Y_i | X_i, a_t, s_t))^2 \right.$$

$$\left. - \left(\frac{1}{T} \sum_{t=1}^{T} (\log p(Y_i | X_i, a_t, s_t)) \right)^2 \right\}.$$

Note that replacement of V_n by $V_n T / (T - 1)$ gives the unbiased estimation of the posterior variance. If a true distribution $q(x, y) = q(x)q(y|x)$ is equal to

$$q(x) = \frac{1}{\sqrt{2\pi}} \exp(-\frac{x^2}{2}),$$

$$q(y|x) = p(y|x, a_0, s_0),$$

then the entropy and the empirical entropy are respectively given by

$$S_n = \frac{1}{2} \log(\pi/2) + \frac{s_0}{2} \sum_{i=1}^{n} (Y_i - a_0 X_i)^2,$$

$$S = \frac{1}{2} \log(\pi/2) + \frac{1}{2}.$$

Remark 15. Let us define the Akaike information criterion in Bayes ($\mathrm{AIC_b}$) and deviance information criterion DIC by the following equations.

$$\mathrm{AIC_b} = T_n + \frac{d}{n},$$

$$\mathrm{DIC} = -\frac{2}{n} \sum_{i=1}^{n} \frac{1}{T} \sum_{t=1}^{T} \log p(Y_i | X_i, a_t, s_t)$$

$$+ \frac{1}{n} \sum_{i=1}^{n} \log p(Y_i | X_i, \bar{a}, \bar{s}),$$

where d is the number or dimension of parameters and $(\bar{a}, \bar{s})$ is the empirical mean of the posterior parameters $\{(a_t \, s_t)\}$. The criteria $\mathrm{AIC_b}$ and DIC can be understood as estimators of the generalization loss. Note that the conventional AIC is defined by using the maximum likelihood estimator, whereas $\mathrm{AIC_b}$ is by the Bayesian predictive density. Hence $\mathrm{AIC} \neq \mathrm{AIC_b}$ in general. If a true distribution is realizable by a statistical model and if the posterior distribution can be approximated by some normal distribution, then AIC, $\mathrm{AIC_b}$, and DIC are the asymptotically unbiased estimators of the generalization loss. If otherwise, the statement does not apply.

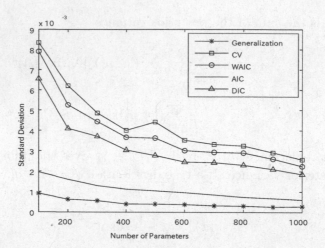

Figure 2.5: Standard deviations caused by posterior sampling. In a simple linear regression problem, standard deviations of the generalization, cross validation, WAIC, AIC$_b$, and DIC errors are compared as a function of the number of the posterior parameters.

Example 7. For a case when $r = 1$, $(a_0, s_0) = (0.3, 0.5)$, and $n = 30$, an experiment was conducted. The true distribution of X was set as the standard normal distribution and that of Y is $a_0 X + \mathcal{N}(0, 1/s_0)$. Firstly, let us examine the fluctuation caused by posterior sampling. The standard deviations of these observables for a fixed sample (X^n, Y^n) in the cases $T = 100, 200, ..., 1000$ are shown in Figure 2.5. The posterior standard deviation of the cross validation errors are larger than other errors. The standard deviations were

$$\sigma(\text{AIC}_b) < \sigma(\text{DIC}) < \sigma(\text{WAIC}) < \sigma(\text{cross validation}).$$

Note that this order is not yet mathematically proved, and it may depend on the conditions about the true distribution and a statistical model. In fact, if the posterior distribution can not be approximated by any normal distribution, the variance of DIC becomes far larger than others. Secondly, we study the averages by comparing them as functions of hyperparameters, $r = -3, -1, 1, 3, 5$. For $r < -1/2$, the prior is improper. However, the posterior is well defined and the cross validation loss and WAIC can estimate the generalization loss. In Figure 2.6 the averages of the generalization error, the cross validation error, WAIC error, AIC$_b$ error, and DIC error are compared. In this experiment their averages were numerically calculated

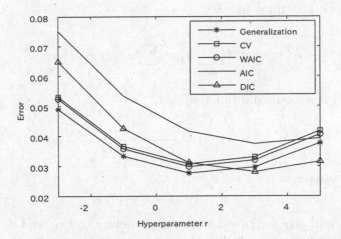

Figure 2.6: Information Criteria as a function of a hyperparameter. In a simple linear regression problem, the averages of the generalization, cross validation, WAIC, AIC_b, and DIC are compared as a function of a hyper parameter r. These are averages over 1000 independent samples with $n = 30$. Both the cross validation error and WAIC error exhibited the same behaviors of the generalization error, whereas neither AIC_b nor DIC did.

using 1000 independent samples of (X^n, Y^n), $n = 30$. As estimators of the generalization error, neither AIC_b nor DIC gave an appropriate function of the hyperparameter, whereas both the cross validation and WAIC did. In the following chapters, we show that the averages of the cross validation loss and WAIC have the same higher order asymptotic behavior as the generalization loss, resulting that they can be employed in evaluation of the average generalization loss as a function of a hyperparameter.

2.4 Neural Network

In many statistical models, the posterior average cannot be calculated analytically, hence the Markov chain Monte Carlo method is necessary (see Chapter 7). Moreover, in statistical models which have hierarchical structures or hidden variables, the posterior distribution cannot be approximated by any normal distribution. An example of such statistical model is an artificial neural network. It is a function $f(x, w) = \{f_j(x, w)\}$ from $x \in \mathbb{R}^M$

to $\mathbb{R}^N$ which is defined by

$$f_j(x, w) = \sigma\Big(\sum_{k=1}^{K} u_{jk}\sigma(\sum_{\ell=1}^{L} w_{k\ell}x_\ell + \theta_k) + \phi_j\Big), \qquad (2.26)$$

where $\sigma(t)$ is a sigmoidal function of t,

$$\sigma(t) = \frac{1}{1 + \exp(-t)}$$

and a parameter w is

$$w = \{u_{jk}, w_{k\ell}, \phi_j, \theta_k\},$$

where u_{jk} and $w_{k\ell}$ are called weight parameters and ϕ_j and θ_k bias parameters. This function is called a three-layer neural network. Recently statistical models which have deeper layers are being applied to many practical problems. The statistical model for a regression problem using a three-layer neural network is a conditional probability density

$$p(y|x, w) = \frac{1}{(2\pi s^2)^{N/2}} \exp\Big(-\frac{1}{2s^2}\|y - f(x, w)\|^2\Big). \qquad (2.27)$$

A neural network can be used also for classification problem. For the case $y \in \{0, 1\}$ its statistical model is represented by

$$p(y|x, w) = f(x, w)^y (1 - f(x, w))^{1-y}.$$

In this model, $f(x, w)$ is used for estimating the conditional probability of y for a given x. By setting a prior on the parameter w, the posterior and predictive distributions are numerically. Since the posterior distribution cannot be approximated by any normal distribution, neither AIC nor BIC can be used for evaluation of a model and a prior. However, WAIC and ISCV can be used.

Example 8. An experiment was conducted about a neural network which had no bias parameters. An input sample $\{x_i \in \mathbb{R}^2; i = 1, 2, ..., n\}$ ($n = 200$) was taken from the uniform distribution in $[-2, 2]^2$. The true conditional distribution was made by $p(y|x, w_0)$ where $p(y|x, w_0)$ was a neural network with three hidden units $H = 3$. A prior was set by the normal distribution $\mathcal{N}(0, 10^2)$ on each u_{jk} and $w_{k\ell}$. The posterior distribution was approximated by a Metropolis method (see Chapter 7.1). We prepared five candidate neural networks which have $H = 1, 2, 3, 4, 5$ hidden units. Figure 2.7 shows the results of 20 trials of:

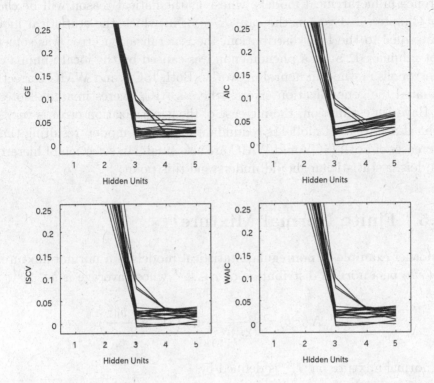

Figure 2.7: Classification problem. Model comparison of neural networks is studied. The true number of the hidden units was three. Both ISCV and WAIC errors correctly estimated the generalization error. AIC overestimated the generalization error, and DIC did not work appropriately.

(1) Upper, left: $G - S$ for $H = 1, 2, 3, 4, 5$
(2) Upper, right: $\mathrm{AIC}_b - S_n$ for $H = 1, 2, 3, 4, 5$
(3) Lower, left: $\mathrm{ISCV} - S_n$ for $H = 1, 2, 3, 4, 5$
(4) Lower, left: $\mathrm{WAIC} - S_n$ for $H = 1, 2, 3, 4, 5$

where G, ISCV, AIC_b, and WAIC are calculated by using the Markov chain Monte Carlo method. In this problem the values of DIC were quite different from others, which are not appropriate for evaluation of hierarchical statistical models. Note that in a neural network the posterior average of the parameter has no meaning. In Bayesian estimation, the generalization errors of a neural network did not so increase even if the statistical model was larger than a true model. This is the general property of Bayesian inference in hierarchical models, whose mathematical reason will be clarified by Chapter 5. Even in the case $H = 3$, in which the statistical model is just equal to the true distribution, the generalization error was sometimes not minimized. Such a phenomenon was caused by the local minima of the Metropolis method in neural networks. Both ISCV and WAIC correctly estimated the generalization errors, whereas AIC_b overestimated. Note that, in Bayesian estimation, the increase of the generalization error is very small even if a statistical model is redundant for a true model, resulting that the increases of both ISCV and WAIC are also small. In selection of hierarchical models, a statistician should understand this point.

2.5 Finite Normal Mixture

Another example of nonregular statistical models is a normal mixture. Let $N(x|b)$ be a normal distribution of $x \in \mathbb{R}^M$ whose average is $b \in \mathbb{R}^M$,

$$N(x|b) = \frac{1}{(2\pi)^{M/2}} \exp\left(-\frac{\|x - b\|^2}{2}\right).$$

A normal mixture on $\mathbb{R}^M$ is defined by

$$p(x|a, b) = \sum_{k=1}^{K} a_k N(x|b_k), \tag{2.28}$$

where $a = (a_1, a_2, ..., a_K)$ and $b = (b_1, b_2, ..., b_K)$ are parameters of a normal mixture, which satisfies $\sum a_j = 1$ and $a_j \geq 0$, and $b_k \in \mathbb{R}^d$. The finite positive integer K is called the number of components. For the prior, we adopt

$$\varphi(a) = \frac{1}{z_1} \prod_{k=1}^{K} (a_k)^{\beta_k - 1},$$

$$\varphi(b) = \frac{1}{z_2} \prod_{k=1}^{K} \exp\left(-\frac{1}{2\sigma^2}\|b_k\|^2\right),$$

where $\varphi(a)$ and $\varphi(b)$ are the Dirichlet distribution with index $\{\beta_k\}$ and the normal distribution, respectively. Here $\beta_k, \sigma^2 > 0$ are hyperparameters and $z_1, z_2 > 0$ are constants.

A variable $y = (y_1, y_2, ..., y_K)$ is competitive if $y_k = 1$ for some k and if $y_\ell = 0$ for other $\ell \neq k$. A statistical model on (x, y) is defined by

$$p(x, y|a, b) = \prod_{k=1}^{K} \Big(a_k N(x|b_k)\Big)^{y_k}. \tag{2.29}$$

Then

$$p(x|a, b) = \sum_y p(x, y|a, b),$$

where the summation is taken over all competitive y. In a normal mixture, by understanding $Y^n = \{Y_i\}$ as hidden or latent variables of a statistical model $p(x, y|a, b)$, the posterior distribution of (a, b, Y^n) is given by

$$p(a, b, Y^n|X^n) \propto \varphi(a)\varphi(b) \prod_{i=1}^{n} p(X_i, Y_i|a, b).$$

By using the Gibbs sampler, which is explained in Chapter 7, we obtain the posterior samples $\{a_t, b_t, Y_t^n\}$. Hence by using $\{a_t, b_t\}$, the posterior and predictive distributions are numerically approximated.

Example 9. Let us study a case $M = 2, n = 100$, where a true distribution was set as:

$$q(x) = \frac{1}{3}N(x|(-2, -2)) + \frac{1}{3}N(x|(0, 0)) + \frac{1}{3}N(x|(2, 2)). \tag{2.30}$$

The hyperparameters of the prior distribution were $\alpha = 0.5$ and $\sigma = 10$. Fifty independent trials were collected and the generalization and cross validation losses were observed. The candidate statistical models were $K = 1, 2, 3, 4, 5$. The posterior distribution was numerically approximated by the Gibbs sampler. We calculated $G - S$, $CV - S_n$, $\text{AIC}_b - S_n$, $DIC - S_n$, and $WAIC - S_n$. Figure 2.8 shows their averages and standard deviations. The generalization loss does not so increase even if the statistical model is redundant for the true distribution. In fact the generalization loss in Figure 2.8 does not increase as AIC. This is the advantage of the Bayesian estimation. However, the increases of the cross validation and WAIC are too small compared to random fluctuations. In practical applications, if the increases of the cross validation and WAIC are far smaller than d/n (d is the number of paramters and n is the sample size) even if the model becomes more complex, then the minimal model in the set of the models which gives almost same cross validation and WAIC should be selected.

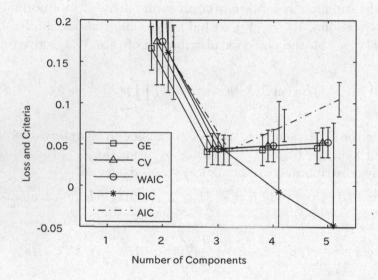

Figure 2.8: Model comparison of normal mixture. Model comparison of a normal mixture is studied. The true density corresponds to $K = 3$. The averages of ISCV and WAIC errors were equal to that of the generalization error, whereas those of DIC and AIC not.

2.6 Nonparametric Mixture

A nonparametric normal mixture is defined by

$$p(x|a,b) = \lim_{K\to\infty}\Big\{\sum_{k=1}^{K} a_k N(x|b_k)\Big\},$$

where $a = (a_1, a_2, ...,)$ and $b = (b_1, b_2, ...)$ are infinite dimensional parameters, which satisfy $\sum a_k = 1$, and $a_k \geq 0$. The prior distributions are set by

$$\varphi(a|\alpha) = \lim_{K\to\infty}\Big\{\frac{1}{z_1(\alpha)} \prod_{k=1}^{K} (a_k)^{\alpha/K-1}\Big\},$$

$$\varphi(b|\sigma) = \lim_{K\to\infty}\Big\{\frac{1}{z_2(\sigma)} \prod_{k=1}^{K} \exp(-\frac{1}{2\sigma^2}\|b_k\|^2)\Big\},$$

where $z_1(\alpha)$ and $z_2(\sigma)$ are normalizing constants. From the mathematical point of view, this model is defined by using Dirichlet process theory [22, 23], by which it is shown that $p(x|a,b)$ is given by the discrete summation with probability one. See Section 10.6.

Although the parameter belongs to the infinite dimensional space, we can construct Markov chain Monte Carlo method by manipulating essentially finite dimentional parameters, resulting that the posterior and predictive distributions are numerically appoximated [21][32]. For example, a Chinese restraurant process [47] and Stick-breaking process [39] are proposed. In [39], it is also proved that nonparametric Bayesian estimation can be accurately approximated by a finite mixture model.

Let $y = (y_1, y_2, ..., y_k, ...)$ be an infinite dimensional competitive variable. Only one k, $y_k = 1$ and others are zero. A statistical model on (x, y) is defined by

$$p(x,y|a,b) = \lim_{K\to\infty}\Big\{\prod_{k=1}^{K}\Big(a_k N(x|b_k)\Big)^{y_k}\Big\}. \tag{2.31}$$

Then

$$p(x|a,b) = \sum_{y} p(x,y|a,b).$$

By using the hidden or latent variable $Y^n = \{Y_i\}$, the posterior distribution of (a, b, Y^n) is given by

$$p(a,b,Y^n|X^n) \propto \varphi(a)\varphi(b) \prod_{i=1}^{n} p(X_i, Y_i|a,b).$$

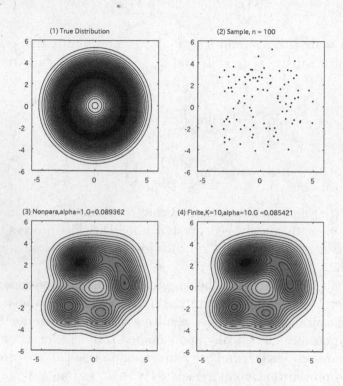

Figure 2.9: Nonparametric and finite mixture. (1) The true distribution cannot be represented by any finite mixture. (2) A sample from the true distribution, $n = 100$. (3) An estimated result by nonparametric Bayes $\alpha = 1$. The generalization error was 0.893. (4) An estimated result by a finite mixture. The generalization error was 0.854. Even if a true distribution is represented by a nonparametric model, the nonparametric method is not always appropriate for statistical estimation.

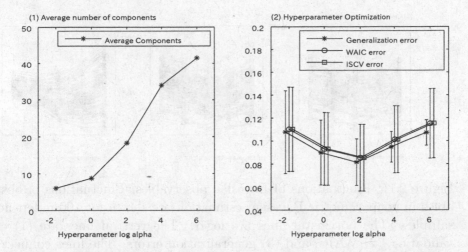

Figure 2.10: Hyperparameter optimization in nonparametric Bayes. (1) The horizontal and vertical lines show the log hyperparameter and the average number of the components. If the hyperparameter increases, then the number of components becomes larger. (2) Generalization error, WAIC error, and ISCV error are compared with respect to the hyperparameter $\log \alpha$.

By using the Gibbs sampler, which is explained in Chapter 7, we obtain the posterior samples $\{a_t, b_t, Y_t^n\}$ which consist of finite dimensional parameters such that the posterior distribution is numerically approximated. Hence we obtain the posterior and predictive densities, resulting that information criteria can be calculated. Sometimes one might think that neither model selection nor hyperparameter optimization could be necessary in the nonparametric method because they should be automatically estimated. However, such consideration is wrong. In general, estimation of something needs its prior. In other words, estimation of a model and a hyperparmeter recursively requires new priors on them. Thus we need the evaluation procedure for preventing the infinite preparation of priors.

Example 10. A true distribution on $x \in \mathbb{R}^2$ was set as

$$q(x) = \lim_{K_0 \to \infty} \left\{ \sum_{k=1}^{K_0} a_{0k} N(x|b_{0k}) \right\},$$

where $N(x|b_{0k})$ is the normal distribution on $\mathbb{R}^2$ and $b_{0k} = (b_{01k}, b_{02k})$ is

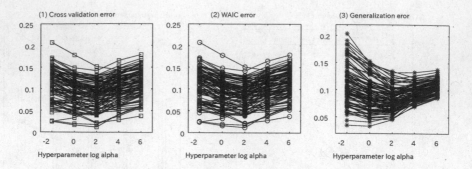

Figure 2.11: Fluctuations of Bayesian observables. Fluctuations of observables in nonparametric Bayesian estimation are shown for 100 independent samples. The horizontal lines are $\log \alpha$. The vertical lines are (1) cross validation, (2) WAIC, and (3) generalization errors. The lines connect the results of the same sample. The optimal α could be found by the cross validation and WAIC. If α was made smaller then the standard deviation of the generalization error became larger.

defined by

$$
\begin{aligned}
a_{0k} &= 1/K_0, \\
b_{01k} &= 3\cos(2\pi k/K_0), \\
b_{02k} &= 3\sin(2\pi k/K_0).
\end{aligned}
$$

The density function of $q(x)$ is shown in Figure 2.9 (1). Note that the true density is not realizable by any finite mixture. (2) A sample $n = 100$ independently taken from $q(x)$ is shown. (3) shows an estimated result by nonparametric method with $\alpha = 1$. The generalization error was 0.0893. (4) An estimated result by a finite mixture with $\alpha = 10$ and $K = 10$ is shown. The generalization error was 0.0824.

If we employ the nonparametric Bayes method, the hyperparameter α should be controlled appropriately, because it strongly affects the estimated result. If α is close to zero, then the average number of components becomes too small. If otherwse, too large. In Figure 2.10 (1), the average number of the components for a given α is shown. One might think that the optimal model selection could be done by the nonparametric method, but such consideration is wrong, because the model selection problem is replaced by the hyperparameter optimization. Also one might think that the hyperparameter α could be optimized by its posterior distribution by preparing its prior, but such consideration is also wrong, because the problem is also replaced by the optimal setting of the hyperprior of α. In other words, the nonparametric Bayes method does not realize the automatic control. On the other hand, even in nonparametric Bayes cases, the optimal hyperparameter can be found by the generalization loss. In Figure 2.10 (2), the average and standard deviation of generalization, cross validation, and WAIC errors are compared for 100 independent trials with $n = 100$. Their fluctuations are shown in Figure 2.11 In this case, the optimal hyperparameter was $\alpha = \exp(2)$, because the true distribution $q(x)$ cannot be realized by any finite mixtures. If the true distribution can be realized by some finite mixtures, then the optimal α would be smaller.

2.7 Problems

1. The predictive density of the normal distribution eq.(2.1) is given by eq.(2.10),

$$p(x|X^n) \ \propto \ \frac{1}{[(x - C_2)^2 + C_1]^{C_3}}, \tag{2.32}$$

where

$$C_1 \ = \ (\hat{\phi}_1\hat{\phi}_3 - \hat{\phi}_2^2)(\hat{\phi}_3 + 1)/\hat{\phi}_3^2, \tag{2.33}$$
$$C_2 \ = \ \hat{\phi}_2/\hat{\phi}_3, \tag{2.34}$$
$$C_3 \ = \ (\hat{\phi}_3 + 1)/2. \tag{2.35}$$

Prove that the average and variance of this predictive distribution are given by

$$\int x\, p(x|X^n)dx \ = \ C_2, \tag{2.36}$$

$$\int (x - C_2)^2 p(x|X^n)dx \ = \ \frac{C_1}{2(C_3 - 3/2)}, \tag{2.37}$$

by using a formula

$$\int_{-\infty}^{\infty} \frac{dx}{[(x - C_2)^2 + C_1]^{C_3}} = \frac{C_1^{1/2 - C_3}\Gamma(C_3 - 1/2)\Gamma(1/2)}{\Gamma(C_3)}.$$

Let us define two estimators of the variance,

$$V_{bayes} = \frac{C_1}{2(C_3 - 3/2)},$$

$$V_{ml} = \frac{1}{n}\sum_{i=1}^{n} X_i^2 - \left(\frac{1}{n}\sum_{i=1}^{n} X_i\right)^2.$$

Prove that, if $\phi_1\phi_3 > \phi_2^2$, then $V_{bayes} > V_{ml}$ holds.

2. In the prior of the multinomial distribution eq.(2.13), let us assume that $\sum_{j=1}^{N} a_j = A$, where $A > 0$ is a constant. Then the cross validation loss C_n in eq.(2.17) as a function of the hyperparameter $a = \{a_j\}$ is minimized if and only if

$$\sum_{j=1}^{N}(n_j/n)\log\frac{(n_j/n)}{(n_j - 1 + a_j)/(n - N + A)} \tag{2.38}$$

is minimized. Also prove that C_n is minimized if and only if

$$a_j = \hat{a}_j, \quad \text{where} \quad \hat{a}_j \equiv 1 + \frac{(A - N)n_j}{n},$$

by using the fact that eq.(2.38) is Kullback-Leibler divergence between two probability distributions. On the other hand, prove that the generalization error using eq.(2.19) and the true entropy,

$$G_n - S = \sum_{j=1}^{N} w_{0j}\log\frac{w_{0j}}{(n_j + a_j)/(n + A)} \tag{2.39}$$

are minimized if and only if $a_j^* = w_{0j}(n + A) - n_j$. Note that $\hat{a}_j$ and a_j^* are the hyperparameters that minimize the cross validation and generalization losses, respectively. The standard deviations of $\hat{a}_j$ and a_j^* are in proportion to $1/\sqrt{n}$ and $\sqrt{n}$, respectively, resulting that $\hat{a}_j \to 1 + (A - N)w_{0j}$ in probability, whereas, a_j^* does not converge even if $n \to \infty$. Note that the random variable $G_n - S$ is different from the average generalization error $\mathbb{E}[G_n] - S$.

3. In the linear regression problem, two approximation methods of eq.(2.25)

for a given function $f(a, s)$ are defined by

$$\mathbb{E}_1[f] = \frac{1}{K}\sum_{k=1}^{K} f(a_k, s_k),$$

$$\mathbb{E}_2[f] = \frac{1}{K}\sum_{k=1}^{K} \int f(a, s_k)p(a|s_k, X^n, Y^n)da,$$

where $\{a_k\}$ and $\{s_k\}$ are independently taken from eq.(2.23) and eq.(2.24) respectively. Prove that $\mathbb{E}_1[f]$ and $\mathbb{E}_2[f]$ have the same average and that the variance of $\mathbb{E}_1[f]$ is not smaller than $\mathbb{E}_2[f]$. In other words, the partial posterior integration makes the variance smaller.

4. Let $f(x, w)$ be a function of a neural network given in eq.(2.26). Then a function $w \mapsto f(x, w)$ is not one-to-one, showing that the posterior distribution does not concentrate on any local parameter region. Therefore, even if the posterior distribution is precisely obtained, the average parameter $\mathbb{E}_w[w]$ is not an appropriate estimator. Discuss the reason why DIC cannot be applied to such statistical models.

5. Let us study model selection problems of a normal mixture given by eq.(2.28). Let $K(n)$ be the optimal number of components in the set $\{1, 2, ..., \infty\}$ that minimizes $\mathbb{E}[G_n]$ for a given sample size n. Discuss the behavior of $K(n)$ if the true distribution is one of the following densities.

$$q_1(x) = \frac{1}{2}N(x|(0.2, 0.2)) + \frac{1}{2}N(x|(0.4, 0.4)),$$

$$q_2(x) = \frac{99}{100}N(x|(-1, -1)) + \frac{1}{100}N(x|(1, 1)),$$

$$q_3(x) = \sum_{k=1}^{\infty} \frac{1}{2^k}N(x|(0.2k, 0.2k)).$$

The probability density functions $q_1(x)$ and $q_2(x)$ consist of two normal distributions. However, they are almost same as one normal distribution, hence $K(n) = 1$ for not so large n. If n is sufficiently large, then $K(n) = 2$. In the case $q_3(x)$, the true distribution is not contained in any finite mixture of normal distributions. In this case $K(n)$ slowly becomes larger when n increases.

Chapter 3

Basic Formula of Bayesian Observables

In this chapter, we introduce the basic Bayesian theory. For an arbitrary triple of a true distribution, a statistical model, and a prior, the behaviors of the free energy or the minus log marginal likelihood, the generalization loss, cross validation loss, training loss, and WAIC are derived by the following procedure.

(1) Firstly, we define the formal relation between a true distribution and a statistical model.

(2) Secondly, definitions of Bayesian observables and their normalized ones are introduced.

(3) Thirdly, the cumulant generating function of the Bayesian prediction is defined.

(4) And lastly, the basic theory of Bayesian statistics is proved by using the cumulant generating function.

At the end of this chapter, we show the recipe for the Bayesian theory construction and its application. In this chaper, we assume that a sample is taken from an unknown true distribution and that a statistical model and a prior are arbitrarily fixed.

3.1 Formal Relation between True and Model

In this section, we define several formal relations between a true probability density $q(x)$ and a statistical model $p(x|w)$.

Definition 4. (Realizability) Let $W \subset \mathbb{R}^d$ be a set of all parameters. If there exists $w_0 \in W$ such that $q(x) = p(x|w_0)$, then $q(x)$ is said to be *realizable* by a statistical model $p(x|w)$. If otherwise, $q(x)$ is *unrealizable*. For a given pair $q(x)$ and $p(x|w)$, the set of true parameters is defined by

$$W_{00} = \{w \in W \; ; \; q(x) = p(x|w) \text{ for arbitrary } x \text{ s.t. } q(x) > 0\}.$$

By the definition, $q(x)$ is realizable by $p(x|w)$ if and only if W_{00} is not the empty set. The set W_{00} is equal to the set of zeros of the Kullback-Leibler distance,

$$W_{00} = \{w \in W \; ; \; \int q(x) \log \frac{q(x)}{p(x|w)} dx = 0\}.$$

If W_{00} is not the empty set, then for an arbitrary $w_0 \in W_{00}$, $p(x|w_0)$ represents the same probability density function $q(x)$. However, derived functions

$$\left(\frac{\partial}{\partial w_j}\right)^k \log p(x|w)\Big|_{w=w_0}$$

may depend on the parameter w_0 in W_{00}.

For a true probability density function $q(x)$ and a statistical model $p(x|w)$, the average log loss function is defined by

$$L(w) = -\int q(x) \log p(x|w) dx. \tag{3.1}$$

It follows that

$$
\begin{aligned}
L(w) &= -\int q(x) \log q(x) dx + \int q(x) \log \frac{q(x)}{p(x|w)} dx \\
&= S + K(q(x)\|p(x|w)),
\end{aligned}
$$

where S is the entropy of the true distribution and $K(q(x)\|p(x|w))$ is the Kullback-Leibler distance from $q(x)$ to $p(x|w)$. If $q(x)$ is realizable by a statistical model, then the average log loss function is minimized if and only if $w \in W_{00}$, and its minimum value is equal to the entropy of the true distribution.

Definition 5. (Regularity) For a given pair of $q(x)$ and $p(x|w)$, let W_0 be the set of minimum points of the average log loss function $L(w)$,

$$W_0 = \{w \in W \; ; \; L(w) = \min_{w'} L(w')\},$$

which is called the set of optimal parameters for the minimum average log loss function. If W_0 consists of a single element w_0 and there exists an open set U such that $w_0 \in U \subset W$ and if the Hessian matrix $\nabla^2 L(w_0)$ at w_0 defined by

$$\left(\nabla^2 L(w_0)\right)_{ij} = \left(\frac{\partial^2 L}{\partial w_i \partial w_j}\right)(w_0) \tag{3.2}$$

is positive definite, then $q(x)$ is said to be *regular* for $p(x|w)$.

By the definition, W_0 is equal to the set of minimum points of the Kullback-Leibler distance $K(q(x)||p(x|w))$. If W is a compact set and if $L(w)$ is a continuous function, then $L(w)$ has a minimum point, hence W_0 is not the empty set. A true probability density $q(x)$ is realizable by a statistical model $p(x|w)$ if and only if

$$W_{00} = W_0.$$

In general, W_0 may contain multiple elements. If a true density is unrealizable by $p(x|w)$, then there may exist $w_1, w_2 \in W_0$ which satisfy $p(x|w_1) \neq p(x|w_2)$.

Definition 6. (Essential uniqueness) Assume that W_0 is not the empty set. If there exists a unique probability density function $p_0(x)$ such that,

$$\text{for arbitrary } w_0 \in W_0, \quad p(x|w_0) = p_0(x),$$

then it is said that the optimal probability density function is *essentially unique*.

If $q(x)$ is realizable by $p(x|w)$, then the optimal probability density function is essentially unique, because $p_0(x) = q(x)$. If W_0 consists of a single element, then the optimal probability density function is unique and essentially unique.

Let us study several cases using examples from the viewpoints of realizability, regularity, and uniqueness.

Example 11. (Realizable, regular, and unique) A true probability density function $q(x)$ and a statistical model $p(x|a)$ are defined by

$$q(x) = \frac{1}{\sqrt{2\pi}} \exp(-\frac{1}{2}x^2),$$

$$p(x|a) = \frac{1}{\sqrt{2\pi}} \exp(-\frac{1}{2}(x-a)^2).$$

In this case, the Kullback-Leibler distance is given by

$$K(a) = \frac{1}{2}a^2.$$

If $W = \{a; -1 \le a \le 1\}$, then

$$W_{00} = W_0 = \{a \; ; \; a = 0\}$$

and the Hessian matrix of the average log loss function is

$$\nabla^2 L(a)|_{a=0} = \nabla^2 K(a)|_{a=0} = 1.$$

Hence $q(x)$ is realizable by and regular for a statistical model $p(x|a)$. If $W = \{a \; ; \; 1 \le a \le 2\}$, then

$$W_{00} = \varnothing, \quad W_0 = \{a \; ; \; a = 1\}.$$

Hence $q(x)$ is unrealizable by and nonregular for a statistical model $p(x|a)$.

Example 12. (Realizable, nonregular, and essentially unique) A true probability density function $q(y|x)q(x)$ and a statistical model $p(y|x, a, b)q(x)$ are defined by

$$q(x) = \begin{cases} 1 & (|x| \le 1) \\ 0 & (|x| > 1) \end{cases},$$

$$q(y|x) = \frac{1}{\sqrt{2\pi}} \exp(-\frac{1}{2}y^2),$$

$$p(y|x, a, b) = \frac{1}{\sqrt{2\pi}} \exp(-\frac{1}{2}(y - a\sin(bx))^2).$$

The set of parameters is defined by

$$W = \{(a, b) \; ; \; |a| \le 1, \; |b| \le \pi/2\}.$$

Then the Kullback-Leibler distance is given by

$$K(a, b) = \frac{a^2}{2} \int_0^1 \sin(bx)^2 dx.$$

Therefore

$$W_{00} = W_0 = \{(a, b) \in W \; ; \; ab = 0\}.$$

The sets W_{00} and W_0 consist of multiple elements. Hence $q(y|x)q(x)$ is realizable by and nonregular for
$p(y|x, a, b)q(x)$. In this case we also say that the conditional density $q(y|x)$ is realizable by and nonregular for $p(y|x, a, b)$.

Example 13. (Unrealizable, regular, and essentially unique) A true probability density function $q(y|x)q(x)$ and a statistical model $p(y|x, a, b)q(x)$ are defined by

$$q(x) = \begin{cases} 1 & (0 \le x \le 1) \\ 0 & \text{(otherwise)} \end{cases},$$

$$q(y|x) = \frac{1}{\sqrt{2\pi}} \exp(-\frac{1}{2}(y - x^2)^2),$$

$$p(y|x, a, b) = \frac{1}{\sqrt{2\pi}} \exp(-\frac{1}{2}(y - ax)^2).$$

The set of all parameters is defined by

$$W = \{(a, b) \; ; \; |a|, |b| \le 1\}.$$

Then the Kullback-Leibler distance is equal to

$$K(a, b) = \frac{1}{2} \int_0^1 (ax - x^2)^2 dx = \frac{a^2}{6} - \frac{a}{4} + \frac{1}{10}.$$

Therefore

$$W_{00} = \varnothing, \quad W_0 = \{a \; ; \; a = 3/4\}.$$

and

$$\nabla^2 L(a)|_{a=3/4} = \nabla^2 K(a)|_{a=3/4} = 1/3.$$

Hence $q(x)q(y|x)$ is unrealizable by and regular for $q(x)p(y|x, a, b)$. Also $q(y|x)$ is unrealizable by and regular for $p(y|x, a, b)$.

Example 14. : (Unrealizable, nonregular, and essentially unique) A true probability density function $q(x, y) = q(x)q(y|x)$ and a statistical model $p(x, y|a, b)$ are defined by

$$q(x, y) = \frac{1}{2\pi} \exp(-\frac{1}{2}\{x^2 + y^2\}),$$

$$p(x, y|a, b) = \frac{1}{2\pi} \exp(-\frac{1}{2}\{(x - 1)^2 + (y - a\sin(bx))^2\}),$$

where the set of all parameters is

$$W = \{(a, b) \; ; \; |a| \le 1, |b| \le \pi/2\}.$$

Then

$$K(a, b) = \frac{1}{2} + \frac{a^2}{2} \int_{-\infty}^{\infty} \sin(bx)^2 q(x) dx.$$

Hence
$$W_{00} = \varnothing, \quad W_0 = \{(a, b) \; ; \; ab = 0\}.$$

It follows that $q(x, y)$ is unrealizable by and nonregular for a statistical model $p(x, y|w)$, however, the optimal model is essentially unique.

Example 15. (Unrealizable, nonregular, and essentially nonunique) A true probability density function $q(x, y)$ and a statistical model $p(x, y|a, b)$ are defined by

$$
\begin{aligned}
q(x, y) &= \frac{1}{2\pi} \exp(-\frac{1}{2}\{x^2 + y^2\}), \\
p(x, y|\theta) &= \frac{1}{2\pi} \exp(-\frac{1}{2}\{(x - \cos\theta)^2 + (y - \sin\theta)^2\}),
\end{aligned}
$$

where the set of all parameters is

$$W = \{\theta \; ; \; -\pi \leq \theta < \pi\}.$$

Then $K(\theta)$ is a constant function of θ,

$$K(\theta) = \frac{1}{2}.$$

Therefore
$$W_{00} = \varnothing, \quad W_0 = \{\theta \; ; \; -\pi \leq \theta < \pi\} = W.$$

For an arbitrary $\theta_0 \in W_0$, the Kullback-Leibler distance $K(q(x)\|p(x|\theta))$ is equal to a constant $1/2$, however, if $\theta_1 \neq \theta_2$ then $p(x, y|\theta_1) \neq p(x, y|\theta_2)$. Therefore $q(x)$ is unrealizable by and nonregular for a statistical model. Moreover, the optimal density is essentially nonunique.

Definition 7. (Relatively finite variance of log density ratio function) Let W and W_0 be sets of parameters and optimal parameters for the minimum average log loss function, respectively. For a given pair $w_0 \in W_0$ and $w \in W$, the log density ratio function is defined by

$$f(x, w_0, w) = \log \frac{p(x|w_0)}{p(x|w)}. \tag{3.3}$$

If there exists $c_0 > 0$ such that, for an arbitrary pair $w_0 \in W_0$ and $w \in W$,

$$\mathbb{E}_X[f(X, w_0, w)] \geq c_0 \mathbb{E}_X[f(X, w_0, w)^2], \tag{3.4}$$

then it is said that the log density ratio function $f(x, w_0, w)$ has a relatively finite variance.

Remark 16. The function $f(x, w_0, w)$ has a relatively finite variance if and only if

$$\sup_{w \notin W_0} \frac{\mathbb{E}_X[f(X, w_0, w)^2]}{\mathbb{E}_X[f(X, w_0, w)]} < \infty. \tag{3.5}$$

If W and W_0 are compact sets and if $\mathbb{E}_X[f(X, w_0, w)]$ and $\mathbb{E}_X[f(X, w_0, w)^2]$ are continuous functions, then both functions have finite values. Hence the condition $w \notin W_0$ in the supremum of eq.(3.5) can be replaced by the condition that w is contained in a neighborhood of of $\mathbb{E}_X[f(X, w_0, w)] = 0$. In other words, the condition $w \notin W_0$ can be replaced by

$$w \in \{w \notin W_0; \mathbb{E}_X[f(X, w_0, w)] < \epsilon\},$$

for an arbitrarily small $\epsilon > 0$.

Example 16. Let us study a case given in Example.12. The log density ratio function is

$$f(x, y, a, b) = -ya \sin(bx) + \frac{a^2 \sin^2(bx)}{2}.$$

Therefore

$$\mathbb{E}_{(X,Y)}[f(x, y, a, b)] = \frac{a^2 b^2}{2} \int_0^1 S^2(bx) x^2 dx,$$

$$\mathbb{E}_{(X,Y)}[f(x, y, a, b)^2] = a^2 b^2 \int_0^1 S^2(bx) x^2 dx + \frac{a^4 b^4}{4} \int_0^1 S^4(bx) x^4 dx,$$

where $S(x) = \sin(x)/x$ with $S(0) = 1$. It follows that, in the neighborhood of $ab = 0$, there exists $c_0 > 0$ such that $\mathbb{E}_{(X,Y)}[f(x, y, a, b)] > c_0 \mathbb{E}_{(X,Y)}[f(x, y, a, b)^2]$. Hence $f(x, y, a, b)$ has a relatively finite variance.

Lemma 3. *Assume that $w_0 \in W_0$ and $w \in W$. If $f(x, w_0, w)$ has a relatively finite variance, then the optimal probability density is essentially unique.*

Proof. Assume that w_1 and w_2 are arbitrary elements of W_0. Then

$$0 = L(w_2) - L(w_1) = \int q(x) f(x, w_1, w_2) dx$$

$$\geq c_0 \int q(x) f(x, w_1, w_2)^2 dx.$$

Hence $f(x, w_1, w_2) = 0$ for an arbitrary x, resulting that $p(x|w_1) - p(x|w_2) = 0$ as a function of x. $\qquad\square$

By Lemma 3, if $f(x, w_0, w)$ has a relatively finite variance, then the optimal probability density function is essentially unique and $f(x, w_0, w)$ does not depend on w_0. In such a case, we use a simple notation

$$f(x, w) = f(x, w_0, w).$$

By the definition of $f(x, w)$, it follows that

$$p(x|w) = p_0(x) \exp(-f(x, w)).$$

Note that if the optimal probability density is not essentially unique, then the log density ratio function does not have a relatively finite variance. The following lemma shows that if a true density is realizable by a statistical model and if the tail probability satisfies a condition, then the log density ratio function has relatively finite variance.

Lemma 4. *Assume that W is a compact set and that $q(x)$ is realizable by $p(x|w)$ and that the log density ratio function $f(x, w) = \log(q(x)/p(x|w))$ is a continuous function of (x, w). If there exists $c_1, c_2 > 0$ such that for an arbitrary $w \in W$,*

$$\int_{|x|>c_1} q(x) f(x, w)^2 dx \le c_2 \int_{|x| \le c_1} q(x) f(x, w)^2 dx,$$

then $f(x, w)$ has a relatively finite variance.

Proof. Since $q(x)$ is realizable by $p(x|w)$, the optimal probability density is uniquely equal to $q(x)$. A function $F(t)$ $(-\infty < t < \infty)$ is defined by

$$F(t) = t + e^{-t} - 1.$$

Then $F'(t) = 1 - e^{-t}$ and $F''(t) = e^{-t} > 0$, resulting that $F(t) \ge 0$ and $F(t) = 0$ if and only if $t = 0$. The constants c_3 and c_4 are defined by

$$c_3 = \sup_{w \in W} \sup_{|x| \le c_1} |f(x, w)|,$$

$$c_4 = \inf_{|t| \le c_3} F(t)/t^2.$$

Then c_3 and c_4 are positive and finite values. Since

$$q(x) F\left(\log \frac{q(x)}{p(x|w)}\right) = q(x) \log \frac{q(x)}{p(x|w)} + p(x|w) - q(x),$$

it follows that

$$\int q(x) \log \frac{q(x)}{p(x|w)} dx = \int q(x) F(f(x,w)) dx$$

$$\geq \int_{|x| \leq c_1} q(x) F(f(x,w)) dx$$

$$\geq c_4 \int_{|x| \leq c_1} q(x) f(x,w)^2 dx$$

$$\geq \frac{c_4}{1 + c_2} \int q(x) f(x,w)^2 dx,$$

which completes the lemma. □

The following lemma shows that, if a true density is regular for a statistical model, then the log density ratio function has relatively finite variance.

Lemma 5. *Assume that W is a compact set and that for an arbitrary pair $w_0 \in W_0$ and $w \in W$, the second derivatives of*

$$\int q(x) f(x, w_0, w) dx, \quad \int q(x) f(x, w_0, w)^2 dx$$

are continuous functions. If $q(x)$ is regular for $p(x|w)$, then the log density ratio function $f(x, w_0, w)$ has a relatively finite variance.

Proof. Since W is a compact set, both functions

$$\int q(x) f(x, w_0, w) dx, \quad \int q(x) f(x, w_0, w)^2 dx$$

have nonnegative values. By the definition,

$$L(w) - L(w_0) = \int q(x) f(x, w_0, w) dx.$$

By the assumption that $q(x)$ is regular for $p(x|w)$, $L(w) - L(w_0) = 0$ if and only if $w = w_0$. Hence it is sufficient to prove that there exists $\epsilon > 0$ such that, in the region $|w - w_0| < \epsilon$,

$$\int q(x) f(x, w_0, w) dx > c_1 \int q(x) f(x, w_0, w)^2 dx$$

for some $c_1 > 0$. By the regularity condition and the mean value theorem, there exists x^* such that

$$L(w) - L(w_0) = \frac{1}{2}(w - w_0)^T \nabla^2 L(w^*)(w - w_0)$$

in a neighborhood of $w = w_0$, hence there exists $\mu_1 > 0$ such that

$$\int q(x)f(x, w_0, w)dx \geq \mu_1 \|w - w_0\|^2$$

in a neighborhood of w_0. On the other hand,

$$\int q(x)f(x, w_0, w)^2 dx$$

is a nonnegative function and is equal to zero at $w = w_0$. Therefore, there exists $\mu_2 > 0$ such that

$$\int q(x)f(X, w_0, w)^2 dx \ \leq \ \mu_2 \|w - w_0\|^2$$

in a neighborhood of w_0. It follows that

$$\sup_{w \notin W_0} \frac{\mathbb{E}_X[f(X, w_0, w)^2]}{\mathbb{E}_X[f(X, w_0, w)]} < \infty,$$

which completes the lemma.   □

Summary Assume that the set of all parameters W is compact. Then the above lemmas show the following relations,

$$\{\text{Regular}\} \ \subset \ \{\text{Relatively Finite Variance}\},$$
$$\{\text{Realizable}\} \ \subset \ \{\text{Relatively Finite Variance}\},$$

and

$$\{\text{Relatively Finite Variance}\} \subset \{\text{Essentially Unique}\}.$$

In this book, we mainly study cases when the log density ratio functions have relatively finite variances. It should be emphasized that such cases include nonregular cases, hence the conventional statistical asymptotic theory does not hold in general.

Example 17. The foregoing examples are classified into the following cases.

- Example 11: Realizable and regular $\rightarrow$ Relatively finite variance.

- Example 12: Realizable $\rightarrow$ Relatively finite variance.

- Example 13: Regular $\rightarrow$ Relatively finite variance.

- Example 14: Nonregular, nonrealizable, but relatively finite variance.

- Example 15: Essentially nonunique → Relatively infinite variance.

In this book, we show that if a log density ratio function has a relatively finite variance, the free energy or the minus log density ratio function F_n, the generalization loss G_n, the cross validation loss C_n, the training loss T_n, and WAIC W_n are subject to the universal statistical laws. That is to say, there exist constants $\lambda, \nu, m > 0$ such that

$$
\begin{aligned}
F_n &= nL_n(w_0) + \lambda \log n + (m-1)O_p(\log \log n) + O_p(1), \\
\mathbb{E}[G_n] &= L(w_0) + \lambda/n + o(1/n), \\
\mathbb{E}[C_n] &= L(w_0) + \lambda/n + o(1/n), \\
\mathbb{E}[W_n] &= L(w_0) + \lambda/n + o(1/n), \\
\mathbb{E}[T_n] &= L(w_0) + (\lambda - 2\nu)/n + o(1/n).
\end{aligned}
$$

Moreover, by defining

$$
L_n(w_0) = -\frac{1}{n} \sum_{i=1}^{n} \log p(X_i|w_0),
$$

the behaviors of random variables satisfy

$$
\begin{aligned}
G_n - L(w_0) + C_n - L_n(w_0) &= 2\lambda/n + o_p(1/n), \\
G_n - L(w_0) + W_n - L_n(w_0) &= 2\lambda/n + o_p(1/n).
\end{aligned}
$$

Note that these mathematical laws hold even if the posterior distribution is quite different from any normal distribution. However, if the log density ratio function does not have a relatively finite variance, then such statistical laws do not hold in general.

3.2 Normalized Observables

In this section, we introduce normalized observables. A triple of a true probability density, a statistical model, and a prior, $(q(x), p(x|w), \varphi(w))$, is fixed.

Let $X^n = (X_1, X_2, ..., X_n)$ be a set of random variables which are independently subject to a true probability density function $q(x)$. Also let X be a random variable which is subject to the same density $q(x)$. Assume that

X and X^n are independent of each other. The average and empirical log loss functions are

$$L(w) \ = \ -\mathbb{E}_X[\log p(X|w)], \tag{3.6}$$

$$L_n(w) \ = \ -\frac{1}{n}\sum_{i=1}^{n}\log p(X_i|w). \tag{3.7}$$

Let W_0 be the set of optimal parameters for the minimum average log loss function $L(w)$. That is to say, W_0 is the set of all parameters which make $L(w)$ smallest. We assume that the log density ratio function of $w_0 \in W_0$ and $w \in W$

$$f(x, w_0, w) = \log\frac{p(x|w_0)}{p(x|w)}$$

has a relatively finite variance. By Lemma 3 the log density ratio function $f(x, w_0, w)$ does not depend on w_0, hence we simply write $f(x, w) = f(x, w_0, w)$. The normalized average and empirical log loss functions are respectively defined by

$$K(w) \ = \ \mathbb{E}_X[f(X, w)], \tag{3.8}$$

$$K_n(w) \ = \ \frac{1}{n}\sum_{i=1}^{n} f(X_i, w). \tag{3.9}$$

By the definition, $-\log p(x|w) = -\log p_0(x) + f(x, w)$, hence

$$L(w) \ = \ L(w_0) + K(w),$$
$$L_n(w) \ = \ L_n(w_0) + K_n(w),$$

and $K(w) \geq 0$. Moreover,

$$K(w) = 0 \Longleftrightarrow w \in W_0.$$

The normalized partition function or the normalized marginal likelihood is defined by

$$Z_n^{(0)} = \int \exp(-nK_n(w))\varphi(w)dw.$$

Then

$$\prod_{i=1}^{n} p(X_i|w) = \left(\prod_{i=1}^{n} p(X_i|w_0)\right)\exp(-nK_n(w))$$

and

$$Z_n = \exp(-nL_n(w_0))\cdot Z_n^{(0)}.$$

Since $L_n(w_0)$ is a constant function of w, the posterior distribution can be rewritten as

$$p(w|X^n) = \frac{1}{Z_n^{(0)}} \exp(-nK_n(w))\varphi(w).$$

The normalized free energy or the normalized minus log marginal likelihood is defined by

$$F_n^{(0)} = -\log \int \exp(-nK_n(w))\varphi(w)dw.$$

The normalized generalization, cross validation, and the training losses and normalized WAIC are also defined by

$$G_n^{(0)} = -\mathbb{E}_X[\log \mathbb{E}_w[\exp(-f(X, w))]],$$

$$C_n^{(0)} = \frac{1}{n}\sum_{i=1}^{n} \log \mathbb{E}_w[\exp(f(X_i, w))],$$

$$T_n^{(0)} = -\frac{1}{n}\sum_{i=1}^{n} \log \mathbb{E}_w[\exp(-f(X_i, w))],$$

$$W_n^{(0)} = T_n^{(0)} + \frac{1}{n}\sum_{i=1}^{n} \mathbb{V}_w[f(X_i, w)],$$

where $\mathbb{E}_w[\ ]$ and $\mathbb{V}_w[\ ]$ are the posterior average and variance, respectively. Here $G_n^{(0)}$, $C_n^{(0)}$, $T_n^{(0)}$, and $W_n^{(0)}$ are sometimes called generalization, cross validation, training, and WAIC errors, respectively.

Lemma 6. *The Bayesian observables and the normalized observables have relations,*

$$\begin{aligned}
F_n &= nL_n(w_0) + F_n^{(0)}, \\
G_n &= L(w_0) + G_n^{(0)}, \\
C_n &= L_n(w_0) + C_n^{(0)}, \\
T_n &= L_n(w_0) + T_n^{(0)}, \\
W_n &= L_n(w_0) + W_n^{(0)}.
\end{aligned}$$

Proof. By the definition and

$$-\log p(x|w) = -\log p_0(x) + f(x, w),$$

this lemma is derived. $\qquad\square$

Remark 17. The Bayesian observables are used in analysis in practical applications, whereas the normalized observables are useful in Bayesian theory construction, because they are mathematically essential quantities.

In the following chapters, we will show that, if a log density ratio function has a relatively finite variance, then $F_n^{(0)}/n$, $G_n^{(0)}$, $C_n^{(0)}$, $T_n^{(0)}$, and $W^{(0)}$ converge to zero in probability, therefore

$$
\begin{aligned}
F_n/n &= L_n(w_0) + O_p(\log n/n), \\
G_n &= L(w_0) + O_p(1/n), \\
C_n &= L_n(w_0) + O_p(1/n), \\
T_n &= L_n(w_0) + O_p(1/n), \\
W_n &= L_n(w_0) + O_p(1/n).
\end{aligned}
$$

By the central limit theorem,

$$
L_n(w_0) - L(w_0) = O_p(1/\sqrt{n}).
$$

Neither $L_n(w_0)$ nor $L(w_p)$ depends on a prior. F_n/n, G_n, C_n, and T_n converge to $L(w_0)$ when $n \to \infty$. If a true distribution is realizable by a statistical model, then $p(x|w_0) = q(x)$ and

$$
\begin{aligned}
F_n/n &= S_n + O_p(\log n/n), \\
G_n &= S + O_p(1/n), \\
C_n &= S_n + O_p(1/n), \\
T_n &= S_n + O_p(1/n), \\
W_n &= S_n + O_p(1/n),
\end{aligned}
$$

where S and S_n are the entropy and the empirical entropy of the true distribution respectively. Neither S nor S_n depends on a statistical model and a prior. Thus the main purpose of the mathematical theory is to clarify the random behaviors of the normalized observables.

3.3   Cumulant Generating Functions

In order to study the asymptotic behaviors of the generalization loss, the cross validation loss, the training loss, and WAIC, the cumulant generating functions are useful.

Definition 8. Let α be a real value. The *cumulant generating functions of generalization and training losses* are respectively defined by

$$\mathcal{G}_n(\alpha) = \mathbb{E}_X[\log \mathbb{E}_w[p(X|w)^\alpha]], \tag{3.10}$$

$$\mathcal{T}_n(\alpha) = \frac{1}{n}\sum_{i=1}^n \log \mathbb{E}_w[p(X_i|w)^\alpha]. \tag{3.11}$$

The kth cumulants are defined by

$$\left(\frac{d}{d\alpha}\right)^k \mathcal{G}_n(0), \quad \left(\frac{d}{d\alpha}\right)^k \mathcal{T}_n(0).$$

Remark 18. Since the definition of the posterior average $\mathbb{E}_w[\ ]$ depends on X^n, the average operations $\mathbb{E}$ and $\mathbb{E}_w$ do not commute, $\mathbb{E}\mathbb{E}_w \neq \mathbb{E}_w\mathbb{E}$. Hence

$$\mathbb{E}[\mathcal{G}_n(\alpha)] \neq \mathbb{E}[\mathcal{T}_n(\alpha)].$$

By the definition,

$$\mathcal{G}_n(0) = \mathcal{T}_n(0) = 0$$

and the generalization, cross validation, training losses and WAIC are given by

$$\begin{aligned}
G_n &= -\mathcal{G}_n(1), \\
C_n &= \mathcal{T}_n(-1), \\
T_n &= -\mathcal{T}_n(1), \\
W_n &= -\mathcal{T}_n(1) + \mathcal{T}_n''(0).
\end{aligned}$$

If we obtain cumulant generating functions as functions of α, then it is easy to calculate the generalization loss, the cross validation loss, the training loss, and WAIC. By using Taylor expansion, the cumulant generating functions are reconstructed by kth cumulants,

$$\mathcal{G}_n(\alpha) = \alpha\mathcal{G}_n'(0) + \frac{\alpha^2}{2}\mathcal{G}_n''(0) + \frac{\alpha^3}{6}\mathcal{G}_n'''(0) + \cdots,$$

$$\mathcal{T}_n(\alpha) = \alpha\mathcal{T}_n'(0) + \frac{\alpha^2}{2}\mathcal{T}_n''(0) + \frac{\alpha^3}{6}\mathcal{T}_n'''(0) + \cdots,$$

if these expansions converge absolutely. Even if these series do not converge absolutely, the asymptotic expansions of $\mathcal{G}_n(\alpha)$ and $\mathcal{T}_n(\alpha)$ can be derived in many cases by the higher order mean value theorem.

Definition 9. Let α be a real value. By using the log density ratio function $f(x, w)$ which satisfies $p(x|w) = p_0(x) \exp(-f(x, w))$, the *normalized cumulant generating functions of generalization and training losses* are respectively defined by

$$\mathcal{G}_n^{(0)}(\alpha) = \mathbb{E}_X[\log \mathbb{E}_w[\exp(-\alpha f(X, w))]],$$

$$\mathcal{T}_n^{(0)}(\alpha) = \frac{1}{n} \sum_{i=1}^{n} \log \mathbb{E}_w[\exp(-\alpha f(X_i, w))].$$

From this definition, simple relations between cumulant generating functions and normalized ones are derived. By using $p(x|w) = p_0(x) \exp(-f(x, w))$ where $p_0(x)$ does not depend on w,

$$\log \mathbb{E}_w[p(X|w)^\alpha] = \alpha \log p_0(X|w) + \log \mathbb{E}_w[\exp(-\alpha f(X, w))].$$

Therefore,

$$\mathcal{G}_n(\alpha) = -\alpha L(w_0) + \mathcal{G}_n^{(0)}(\alpha), \qquad (3.12)$$

$$\mathcal{T}_n(\alpha) = -\alpha L_n(w_0) + \mathcal{T}_n^{(0)}(\alpha). \qquad (3.13)$$

Hence, for $k = 0, 2, 3, 4, \dots$ $(k \neq 1)$ the kth cumulants satisfy

$$\left(\frac{d}{d\alpha}\right)^k \mathcal{G}_n(0) = \left(\frac{d}{d\alpha}\right)^k \mathcal{G}_n^{(0)}(0),$$

$$\left(\frac{d}{d\alpha}\right)^k \mathcal{T}_n(0) = \left(\frac{d}{d\alpha}\right)^k \mathcal{T}_n^{(0)}(0).$$

For $k = 1$,

$$\left(\frac{d}{d\alpha}\right) \mathcal{G}_n(0) = -L(w_0) + \left(\frac{d}{d\alpha}\right) \mathcal{G}_n^{(0)}(0),$$

$$\left(\frac{d}{d\alpha}\right) \mathcal{T}_n(0) = -L_n(w_0) + \left(\frac{d}{d\alpha}\right) \mathcal{T}_n^{(0)}(0).$$

Definition 10. Let α be a real value. For a given random variable A, we use notations,

$$\ell_k(A) = \frac{\mathbb{E}_w[(\log p(A|w))^k p(A|w)^\alpha]}{\mathbb{E}_w[p(A|w)^\alpha]},$$

$$\ell_k^{(0)}(A) = \frac{\mathbb{E}_w[(-f(A, w))^k \exp(-\alpha f(A, w))]}{\mathbb{E}_w[\exp(-\alpha f(A, w))]}.$$

If $\alpha = 0$, these are equal to the posterior averages of kth power,

$$\ell_k(A) = \mathbb{E}_w[(\log p(A|w))^k],$$

$$\ell_k^{(0)}(A) = \mathbb{E}_w[(-f(A, w))^k].$$

Lemma 7. *The cumulant generating functions satisfy*

$$
\begin{aligned}
\mathcal{G}'_n(\alpha) &= \mathbb{E}_X[\ell_1(X)], \\
\mathcal{G}''_n(\alpha) &= \mathbb{E}_X[\ell_2(X) - \ell_1(X)^2], \\
\mathcal{T}'_n(\alpha) &= \frac{1}{n}\sum_{i=1}^{n}\{\ell_1(X_i)\}, \\
\mathcal{T}''_n(\alpha) &= \frac{1}{n}\sum_{i=1}^{n}\{\ell_2(X_i) - \ell_1(X_i)^2\}.
\end{aligned}
$$

The same equations hold for normalized functions,

$$
\begin{aligned}
\mathcal{G}_n^{(0)\prime}(\alpha) &= \mathbb{E}_X[\ell_1^{(0)}(X)], \\
\mathcal{G}_n^{(0)\prime\prime}(\alpha) &= \mathbb{E}_X[\ell_2^{(0)}(X) - \ell_1^{(0)}(X)^2], \\
\mathcal{T}_n^{(0)\prime}(\alpha) &= \frac{1}{n}\sum_{i=1}^{n}\{\ell_1^{(0)}(X_i)\}, \\
\mathcal{T}_n^{(0)\prime\prime}(\alpha) &= \frac{1}{n}\sum_{i=1}^{n}\{\ell_2^{(0)}(X_i) - \ell_1^{(0)}(X_i)^2\}.
\end{aligned}
$$

Proof. For an arbitrary function $g(\alpha)$, let $\partial^k g(\alpha)$ be the kth order derived function of $g(\alpha)$. Then

$$
\partial \log g(\alpha) = \frac{\partial g(\alpha)}{g(\alpha)},
$$

and

$$
\partial\left(\frac{\partial^k g(\alpha)}{g(\alpha)}\right) = \left(\frac{\partial^{k+1} g(\alpha)}{g(\alpha)}\right) - \left(\frac{\partial^k g(\alpha)}{g(\alpha)}\right)\left(\frac{\partial g(\alpha)}{g(\alpha)}\right).
$$

By using this equation recursively, the lemma is obtained. $\qquad\square$

Remark 19. By the same method, the higher order cumulants can be derived. For example,

$$
\begin{aligned}
\mathcal{G}'''_n(\alpha) &= \mathbb{E}_X[\ell_3(X) - 3\ell_2(X)\ell_1(X) + 2\ell_1(X)^3], \\
\mathcal{G}''''_n(\alpha) &= \mathbb{E}_X[\ell_4(X) - 4\ell_3(X)\ell_1(X) - 3\ell_2(X)^2 \\
&\quad + 12\ell_2(X)\ell_1(X)^2 - 6\ell_1(X)^4]. \\
\mathcal{T}'''_n(\alpha) &= \frac{1}{n}\sum_{i=1}^{n}\{\ell_3(X_i) - 3\ell_2(X_i)\ell_1(X_i) + 2\ell_1(X_i)^3\}, \\
\mathcal{T}''''_n(\alpha) &= \frac{1}{n}\sum_{i=1}^{n}\{\ell_4(X_i) - 4\ell_3(X_i)\ell_1(X_i) - 3\ell_2(X)^2 \\
&\quad + 12\ell_2(X_i)\ell_1(X_i)^2 - 6\ell_1(X_i)^4\}.
\end{aligned}
$$

For the normalized case, the same equations hold by replacing $\ell_k(A)$ by $\ell_k^{(0)}(A)$.

By using these equations, cumulants are given by follows. For $k = 1$.

$$
\begin{aligned}
\mathcal{G}'(0) &= \mathbb{E}_X[\mathbb{E}_w[\log p(X|w)]] \\
&= -L(w_0) - \mathbb{E}_w[K(w)], \tag{3.14} \\
\mathcal{T}'(0) &= \frac{1}{n}\sum_{i=1}^{n}\mathbb{E}_w[\log p(X_i|w)] \\
&= -L_n(w_0) - \mathbb{E}_w[K_n(w)]. \tag{3.15}
\end{aligned}
$$

For $k = 2$,

$$
\begin{aligned}
\mathcal{G}''(0) &= \mathbb{E}_X[\mathbb{E}_w[(\log p(X|w))^2] - \mathbb{E}_w[\log p(X|w)]^2] \\
&= \mathbb{E}_X[\mathbb{E}_w[f(X,w)^2] - \mathbb{E}_w[f(X,w)]^2], \\
&= \mathbb{E}_X[\mathbb{V}_w[f(X,w)]], \tag{3.16} \\
\mathcal{T}''(0) &= \frac{1}{n}\sum_{i=1}^{n}\{\mathbb{E}_w[\log p(X_i|w))^2] - \mathbb{E}_w[\log p(X_i|w)]^2\} \\
&= \frac{1}{n}\sum_{i=1}^{n}\{\mathbb{E}_w[f(X_i,w)^2] - \mathbb{E}_w[f(X_i,w)]^2\} \\
&= \frac{1}{n}\sum_{i=1}^{n}\mathbb{V}_w[f(X_i,w)], \tag{3.17}
\end{aligned}
$$

where $\mathbb{V}_w[f(w)]$ is the variance of $f(w)$ in the posterior distribution.

Lemma 8. *Let $c_2 = 2$, $c_3 = 6$, $c_4 = 26$. Then for $k = 2, 3, 4$,*

$$
\left|\left(\frac{d}{d\alpha}\right)^k \mathcal{G}_n(\alpha)\right| \leq c_k \mathbb{E}_X\left[\frac{\mathbb{E}_w[|f(X,w)|^k \exp(-\alpha f(X,w))]}{\mathbb{E}_w[\exp(-\alpha f(X,w))]}\right],
$$

$$
\left|\left(\frac{d}{d\alpha}\right)^k \mathcal{T}_n(\alpha)\right| \leq c_k \frac{1}{n}\sum_{i=1}^{n}\frac{\mathbb{E}_w[|f(X_i,w)|^k \exp(-\alpha f(X_i,w))]}{\mathbb{E}_w[\exp(-\alpha f(X_i,w))]}.
$$

Proof. For an arbitrary random variable A and an arbitrary function $g(A, w)$, an expectation operator $\mathbb{E}_w^{(\alpha)}[\ ]$ is defined by

$$
\mathbb{E}_w^{(\alpha)}[g(A,w)] \equiv \frac{\mathbb{E}_w[g(A,w)\exp(-\alpha f(A,w))]}{\mathbb{E}_w[\exp(-\alpha f(A,w))]}.
$$

Let us prove the first inequality for $k = 3$.

$$
\left(\frac{d}{d\alpha}\right)^3 \mathcal{G}_n(\alpha) = \mathbb{E}_X\Big[\mathbb{E}_w^{(\alpha)}[f(X,w)^3]
$$
$$
- 3\mathbb{E}_w^{(\alpha)}[f(X,w)^2]\mathbb{E}_w^{(\alpha)}[f(X,w)] + 2\mathbb{E}_w^{(\alpha)}[f(X,w)]^3\Big].
$$

Then by using Hölder's inequality, for arbitrary $1 \leq j \leq k$,

$$
\mathbb{E}_w^{(\alpha)}[|f(A,w)|^j] \leq \mathbb{E}_w^{(\alpha)}[|f(A,w)|^k]^{j/k}.
$$

By applying this inequality, it follows that

$$
\left|\left(\frac{d}{d\alpha}\right)^3 \mathcal{G}_n(\alpha)\right| \leq 6\mathbb{E}_X\Big[\mathbb{E}_w^{(\alpha)}[|f(x,w)|^3]\Big].
$$

For the other cases, the same method can be applied, which completes the lemma. $\qquad\square$

3.4 Basic Bayesian Theory

By combining the foregoing observables, we obtain the basic theorem of Bayesian statistics.

Theorem 3. *(Basic theorem) Let $\mathcal{G}'(0)$, $\mathcal{T}'(0)$, $\mathcal{G}''(0)$, and $\mathcal{T}''(0)$ be random variables defined by eq.(3.14), ...,eq.(3.17). Assume that*

$$
\sup_{|\alpha|\leq 1}\left|\left(\frac{d}{d\alpha}\right)^3 \mathcal{G}_n(\alpha)\right| = o_p(\frac{1}{n}), \tag{3.18}
$$

$$
\sup_{|\alpha|\leq 1}\left|\left(\frac{d}{d\alpha}\right)^3 \mathcal{T}_n(\alpha)\right| = o_p(\frac{1}{n}). \tag{3.19}
$$

Then the generalization loss, the cross validation loss, the training loss, and WAIC are given by

$$
G_n = -\mathcal{G}_n(1) = -\mathcal{G}_n'(0) - \frac{1}{2}\mathcal{G}_n''(0) + o_p(\frac{1}{n}), \tag{3.20}
$$

$$
T_n = -\mathcal{T}_n(1) = -\mathcal{T}_n'(0) - \frac{1}{2}\mathcal{T}_n''(0) + o_p(\frac{1}{n}). \tag{3.21}
$$

$$
C_n = \mathcal{T}_n(-1) = -\mathcal{T}_n'(0) + \frac{1}{2}\mathcal{T}_n''(0) + o_p(\frac{1}{n}). \tag{3.22}
$$

$$
W_n = -\mathcal{T}_n(1) + \mathcal{T}_n''(0) = -\mathcal{T}_n'(0) + \frac{1}{2}\mathcal{T}_n''(0) + o_p(\frac{1}{n}). \tag{3.23}
$$

Assume that

$$\mathbb{E}[\sup_{|\alpha|\leq 1}\left|\left(\frac{d}{d\alpha}\right)^3 \mathcal{G}_n(\alpha)\right|] = o(\frac{1}{n}), \qquad (3.24)$$

$$\mathbb{E}[\sup_{|\alpha|\leq 1}\left|\left(\frac{d}{d\alpha}\right)^3 \mathcal{T}_n(\alpha)\right|] = o(\frac{1}{n}). \qquad (3.25)$$

Then the averages of the generalization loss, the cross validation loss, the training loss, and WAIC are given by

$$\mathbb{E}[G_n] = -\mathbb{E}[\mathcal{G}'_n(0)] - \frac{1}{2}\mathbb{E}[\mathcal{G}''_n(0)] + o(\frac{1}{n}), \qquad (3.26)$$

$$\mathbb{E}[T_n] = -\mathbb{E}[\mathcal{T}'_n(0)] - \frac{1}{2}\mathbb{E}[\mathcal{T}''_n(0)] + o(\frac{1}{n}). \qquad (3.27)$$

$$\mathbb{E}[C_n] = -\mathbb{E}[\mathcal{T}'_n(0)] + \frac{1}{2}\mathbb{E}[\mathcal{T}''_n(0)] + o(\frac{1}{n}), \qquad (3.28)$$

$$\mathbb{E}[W_n] = -\mathbb{E}[\mathcal{T}'_n(0)] + \frac{1}{2}\mathbb{E}[\mathcal{T}''_n(0)] + o(\frac{1}{n}). \qquad (3.29)$$

Proof. By using the mean value theorem, for a given α there exists α^* such that $|\alpha^*| \leq |\alpha|$ and that

$$\mathcal{G}_n(\alpha) = \mathcal{G}_n(0) + \alpha\mathcal{G}'_n(0) + \frac{1}{2}\alpha^2\mathcal{G}''_n(0) + \frac{1}{6}\alpha^3\mathcal{G}_n^{(3)}(\alpha^*).$$

The case $\alpha = 1$ gives the first half of the theorem. The latter half is derived by the same method for $\mathcal{T}_n(\alpha)$ and $\alpha = \pm 1$. $\square$

The conditions given by eqs.(3.18), (3.19), (3.24), and (3.25) are proved in the following chapters for several circumstances. Moreover, we will prove that if the log density ratio function has a relatively finite variance, there exist constants $\lambda, \nu > 0$ such that

$$\mathcal{G}_n^{(0)\prime}(0) + \mathcal{T}_n^{(0)\prime}(0) = -\frac{2\lambda}{n} + o_p(\frac{1}{n}),$$

$$\mathcal{G}_n^{(0)\prime\prime}(0) = \frac{\nu}{n} + o_p(\frac{1}{n}),$$

$$\mathcal{T}_n^{(0)\prime\prime}(0) = \frac{\nu}{n} + o_p(\frac{1}{n}),$$

and that their averages satisfy the equation

$$\mathbb{E}\left[\mathcal{G}_n^{(0)\prime}(0) + \frac{\mathcal{G}_n^{(0)\prime\prime}(0)}{2}\right] = \mathbb{E}\left[\mathcal{T}_n^{(0)\prime}(0) - \frac{\mathcal{T}_n^{(0)\prime\prime}(0)}{2}\right] + o(\frac{1}{n}).$$

It is very important that these equations hold even if the posterior distribution cannot be approximated by any normal distribution. These equations are universal laws in Bayesian statistics. Also these equations except eq.(3.22) and eq.(3.28) hold even if a sample consists of conditionally independent random variables. Note that if a sample consists of independent random variables, then

$$\mathbb{E}[C_n] = \mathbb{E}[G_{n-1}].$$

However, this equation does hold without independency. The cross validation can be employed in the case when a sample is independent.

Based on the above theorem, for a given triple $(q(x), p(x|w), \varphi(w))$, Bayesian theory can be derived by the following recipe. The theoretical behaviors of the free energy or the minus log marginal likelihood, the generalization loss, the cross validation loss, the training losses, and WAIC are clarified by the following procedures.

Recipe for Bayesian Theory Construction

1. An arbitrary triple of a true distribution, a statistical model, and a prior $(q(x), p(x|w), \varphi(w))$ is chosen and fixed. The set of all parameters is denoted by W. Assume that X^n is a sample which consists of independent random variables subject to $q(x)$.

2. The empirical and average log loss functions are defined by

$$L_n(w) = -\frac{1}{n}\sum_{i=1}^{n}\log p(X_i|w),$$

$$L(w) = -\int q(x)\log p(x|w)dx.$$

Find the set of optimal parameters which minimize $L(w)$,

$$W_0 = \{w \in W \; ; \; L(w) = \min_{w' \in W} L(w')\}.$$

3. Check that the log density ratio function made of $q(x)$ and $p(x|w)$ has a relatively finite variance. Then

$$f(x, w) = \log \frac{p(x|w_0)}{p(x|w)}$$

does not depend on a choice of $w_0 \in W_0$.

4. Define the average and empirical log likelihood ratio functions

$$K(w) = \int f(x, w)q(x)dx,$$

$$K_n(w) = \frac{1}{n}\sum_{i=1}^{n} f(X_i, w).$$

The normalized partition function or the normalized marginal likelihood is given by

$$Z_n^{(0)} = \int \exp(-nK_n(w))\varphi(w)dw.$$

Then the free energy or the minus log marginal likelihood is given by

$$F_n = nL_n(w_0) - \log Z_n^{(0)}.$$

5. The average by the posterior distribution is equal to

$$\mathbb{E}_w[\ ] = \frac{\int(\ \ )\exp(-nK_n(w))\varphi(w)dw}{\int \exp(-nK_n(w))\varphi(w)dw}.$$

Calculate $\mathbb{E}_w[f(x, w)]$ and $\mathbb{V}_w[f(x, w)]$. Then

$$\mathbb{E}_w[K(w)] = \mathbb{E}_X\mathbb{E}_w[f(X, w)],$$

$$\mathbb{E}_w[K_n(w)] = \frac{1}{n}\sum_{i=1}^{n} \mathbb{E}_w[f(X_i, w)],$$

$$\mathbb{E}_X\mathbb{V}_w[f(X, w)], \quad \frac{1}{n}\sum_{i=1}^{n} \mathbb{V}_w[f(X_i, w)]$$

are obtained.

6. Based on the basic Theorem 3, the generalization, cross validation, and training losses are given by

$$G_n = L(w_0) + \mathbb{E}_w[K(w)] - \frac{1}{2}\mathbb{E}_X\mathbb{V}_w[f(X, w)] + o_p(\frac{1}{n}),$$

$$C_n = L_n(w_0) + \mathbb{E}_w[K_n(w)] + \frac{1}{2n}\sum_{i=1}^{n} \mathbb{V}_w[f(X_i, w)] + o_p(\frac{1}{n}),$$

$$T_n = L_n(w_0) + \mathbb{E}_w[K_n(w)] - \frac{1}{2n}\sum_{i=1}^{n} \mathbb{V}_w[f(X_i, w)] + o_p(\frac{1}{n}).$$

Note that WAIC, W_n, has the same expansion as C_n.

Example 18. Let us apply the recipe to a simple case.

1. Let $A > 0$ be a constant and $W = \{w \in \mathbb{R}; |w| \leq A\}$. Let us derive the Bayesian statistical theory for the case when a triple $(q(x), p(x|w), \varphi(w))$ is given by

$$q(x) = \frac{1}{\sqrt{2\pi\sigma^2}} \exp\left(-\frac{x^2}{2\sigma^2}\right),$$

$$p(x|w) = \frac{1}{\sqrt{2\pi}} \exp\left(-\frac{(x-w)^2}{2}\right),$$

$$\varphi(w) = \frac{1}{2A},$$

where $\sigma^2 > 0$ is not a parameter but a constant. Note that this true distribution is realizable by a statistical model if and only if $\sigma^2 = 1$.

2. The empirical and average log loss functions are respectively given by

$$L_n(w) = \frac{1}{2}\log(2\pi) + \frac{1}{2}(\sigma_n^2 + w^2 - \frac{2}{\sqrt{n}}w\xi_n),$$

$$L(w) = \frac{1}{2}\log(2\pi) + \frac{1}{2}(\sigma^2 + w^2),$$

where

$$\sigma_n^2 = \frac{1}{n}\sum_{i=1}^{n} X_i^2,$$

$$\xi_n = \frac{1}{\sqrt{n}}\sum_{i=1}^{n} X_i.$$

The random variance σ_n^2 converges to σ^2 in probability. The random variable ξ_n is subject to the normal distribution whose average and variance are zero and σ^2 respectively. The average log loss function $L(w)$ is minimized if and only if $w = 0$, resulting that $W_0 = \{0\}$.

3. The log density ratio function is

$$f(x, w) = \log\frac{p(x|w_0)}{p(x|w)} = \frac{w^2}{2} - wx,$$

where we used $w_0 = 0$. It follows that

$$\mathbb{E}_X[f(X, w)] = \frac{w^2}{2},$$

$$\mathbb{E}_X[f(X, w)^2] = \frac{w^4}{4} + w^2\sigma^2.$$

Thus $f(x, w)$ has a relatively finite variance by using eq.(3.5) and $|w| \leq A$.

4. The average and empirical log likelihood ratio functions are respectively given by

$$K(w) = \int f(x, w)q(x)dx = \frac{w^2}{2},$$

$$K_n(w) = \frac{1}{n}\sum_{i=1}^{n} f(X_i, w) = \frac{w^2}{2} - \frac{w\xi_n}{\sqrt{n}}.$$

The normalized partition function or the normalized marginal likelihood is equal to

$$\begin{aligned}
Z_n^{(0)} &= \frac{1}{2A}\int_{-A}^{A} \exp(-\frac{n\,w^2}{2} + \sqrt{n}\,w\,\xi_n)dw \\
&= \frac{1}{2A}\int_{-\infty}^{\infty} \exp(-\frac{n(w - \xi_n/\sqrt{n})^2}{2} + \frac{\xi_n^2}{2})dw + O_p(e^{-n}) \\
&= \frac{\sqrt{2\pi}}{2A\sqrt{n}} \exp(\xi_n^2/2) + O_p(e^{-n}).
\end{aligned}$$

Therefore the theoretical behavior of the free energy or the minus log marginal likelihood is derived as

$$\begin{aligned}
F_n &= nL_n(w_0) - \log Z_n^{(0)} \\
&= \frac{n-1}{2}\log(2\pi) + \frac{n\sigma_n^2 - \xi_n^2}{2} + \frac{1}{2}\log n + \log(2A) + o_p(1).
\end{aligned}$$

Its average is given by

$$\mathbb{E}[F_n] = \frac{n-1}{2}\log(2\pi) + \frac{1}{2}(n\sigma^2 + \log n - 1) + \log(2A) + o(1).$$

5. The above calculation for $Z_n^{(0)}$ shows that the posterior distribution is asymptotically given by

$$p(w|X^n) = \sqrt{\frac{n}{2\pi}}\exp\left(-\frac{n}{2}(w - \frac{\xi_n}{\sqrt{n}})^2\right) + O_p(e^{-n}).$$

Let $\mathbb{E}_w^*[\ ]$ and $\mathbb{V}^*[\ ]$ be the average and variance operators using the normal distribution whose average and variance are $\xi_n/\sqrt{n}$ and $1/n$, respectively. Then

$$\mathbb{E}_w^*[w] = \frac{\xi_n}{\sqrt{n}},$$

$$\mathbb{E}_w^*[w^2] = \frac{1 + \xi_n^2}{n},$$

$$\mathbb{E}_w^*[w^k] = O_p(\frac{1}{n^{k/2}}) \quad (k \geq 3),$$

resulting that

$$\mathbb{E}_w[f(x, w)] = \mathbb{E}_w^*[f(x, w)] + O_p(e^{-n})$$

$$= \frac{1 + \xi_n^2}{2n} - \frac{x\xi_n}{\sqrt{n}} + o_p(1/n),$$

$$\mathbb{V}_w[f(x, w)] = \mathbb{V}_w^*[f(x, w)] + O_p(e^{-n}) = \frac{x^2}{n} + O_p(1/n^{3/2}).$$

Hence

$$\mathbb{E}_w[K(w)] = \mathbb{E}_X \mathbb{E}_w[f(X, w)] = \frac{1 + \xi_n^2}{2n} + o_p(1/n),$$

$$\mathbb{E}_w[K_n(w)] = \frac{1}{n} \sum_{i=1}^{n} \mathbb{E}_w[f(X_i, w)] = \frac{1 - \xi_n^2}{2n} + o_p(1/n),$$

$$\mathbb{E}_X \mathbb{V}_w[f(X, w)] = \frac{\sigma^2}{n} + o_p(1/n),$$

$$\frac{1}{n} \sum_{i=1}^{n} \mathbb{V}_w[f(X_i, w)] = \frac{\sigma_n^2}{n} + o_p(1/n).$$

6. By using the results above, and $\sigma_n^2 = \sigma^2 + O_p(1/\sqrt{n})$, we obtain the Bayesian statistical theory,

$$G_n = \frac{1}{2} \log(2\pi) + \frac{\sigma^2}{2} + \frac{1 + \xi_n^2 - \sigma^2}{2n} + o_p(\frac{1}{n}),$$

$$C_n = \frac{1}{2} \log(2\pi) + \frac{\sigma_n^2}{2} + \frac{1 - \xi_n^2 + \sigma^2}{2n} + o_p(\frac{1}{n}),$$

$$T_n = \frac{1}{2} \log(2\pi) + \frac{\sigma_n^2}{2} + \frac{1 - \xi_n^2 - \sigma^2}{2n} + o_p(\frac{1}{n}).$$

Note that

$$(G_n - L(w_0)) + (C_n - L_n(w_0)) = \frac{1}{n} + o_p(1/n)$$

holds. Since $\mathbb{E}[\xi_n^2] = \sigma^2$, their averages are

$$
\begin{aligned}
\mathbb{E}[G_n] &= \frac{1}{2}\log(2\pi) + \frac{\sigma^2}{2} + \frac{1}{2n} + o(\frac{1}{n}), \\
\mathbb{E}[C_n] &= \frac{1}{2}\log(2\pi) + \frac{\sigma^2}{2} + \frac{1}{2n} + o(\frac{1}{n}), \\
\mathbb{E}[T_n] &= \frac{1}{2}\log(2\pi) + \frac{\sigma^2}{2} + \frac{1-2\sigma^2}{2n} + o(\frac{1}{n}).
\end{aligned}
$$

The random variable WAIC and its expected value, W_n and $\mathbb{E}[W_n]$, have the same asymptotic expansions as C_n and $\mathbb{E}[C_n]$, respectively. The average generalization loss is a decreasing function of n, whereas the average training loss is increasing. In the following chapters, we show that if a log density ratio function has a relatively finite variance, then the generalization loss, the cross validation loss, the training loss, and WAIC have the same asymptotic behaviors as this case.

Example 19. If a log density ratio function does not have a relatively finite variance, then asymptotic behaviors are different in general. Let us study the case given in Example 15. In this case, we cannot employ the above recipe for theory construction. Let (X_i, Y_i) $(i = 1, 2, ..., n)$ be pairs of random variables which are subject to $q(x, y)$ in Example 15. We use notations,

$$
\begin{aligned}
r_n^2 &= \frac{1}{n}\sum_{i=1}^{n}\{X_i^2 + Y_i^2\}, \\
\xi_n &= \frac{1}{\sqrt{n}}\sum_{i=1}^{n}X_i, \\
\eta_m &= \frac{1}{\sqrt{n}}\sum_{i=1}^{n}Y_i.
\end{aligned}
$$

The empirical entropy is

$$
S_n = -\frac{1}{n}\sum_{i=1}^{n}\log q(X_i, Y_i) = \log(2\pi) + \frac{r_n^2}{2}.
$$

The partition function is

$$
Z_n = \int_{-\pi}^{\pi} \frac{1}{(2\pi)^n} \exp(-\frac{1}{2}\sum_{i=1}^{n}\{(X_i - \cos\theta)^2 + (Y_i - \sin\theta)^2\})\frac{d\theta}{2\pi}
$$

$$
= \frac{\exp(-n(r_n^2+1)/2)}{(2\pi)^{n+1}} \int_{-\pi}^{\pi} \exp(\sqrt{n}\xi_n\cos\theta + \sqrt{n}\eta_n\sin\theta)d\theta
$$

$$
= \frac{\exp(-n(r_n^2+1)/2 + \sqrt{n}\gamma_n)}{(2\pi)^{n+1}} \int_{-\pi}^{\pi} \exp(\sqrt{n}\gamma_n(\cos\theta - 1)))d\theta
$$

where we used $\gamma_n = \sqrt{\xi_n^2 + \eta_n^2}$ and the cyclic condition of θ. If n is sufficiently large, then the main part of integration $[-\pi, \pi]$ is the neighborhood of the origin. By using

$$
1 - \cos\theta = -\theta^2/2 + O(\theta^4),
$$

and

$$
\int_{-\pi}^{\pi} \exp(-\sqrt{n}\gamma_n\theta^2/2)d\theta = \sqrt{\frac{2\pi}{n^{1/2}\gamma_n}} + o_p(\exp(-\sqrt{n}\gamma_n)),
$$

The free energy is

$$
F_n = n(r_n^2+1)/2 - \sqrt{n}\gamma_n + (n+1/2)\log(2\pi)
$$
$$
+ \frac{1}{4}\log n + \log\gamma_n + o_p(1).
$$

Since both ξ_n and y_n are subject to the normal distribution with average 0 and variance 1, γ_n^2 is subject to the chi-squared distribution with 2 degrees of the freedom. Therefore,

$$
\mathbb{E}[\gamma_n] = \frac{1}{2\Gamma(1)} \int_0^{\infty} x^{1/2}e^{-x/2}dx = \sqrt{2}\Gamma(3/2),
$$

$$
\mathbb{E}[\gamma_n^2] = \frac{1}{2\Gamma(1)} \int_0^{\infty} xe^{-x/2}dx = 2.
$$

The average free energy is given by

$$
\mathbb{E}[F_n] = (3/2 + \log(2\pi))n - \sqrt{2}\Gamma(3/2)\sqrt{n} + \frac{1}{4}\log n + O(1).
$$

By Remark 14 and $\mathbb{E}[F_n] = \mathbb{E}[G_{n+1}] - \mathbb{E}[G_n]$, if $\mathbb{E}[G_n]$ has an asymptotic expansion, then it is equal to

$$
\mathbb{E}[G_n] = (3/2 + \log(2\pi)) - \frac{\Gamma(3/2)}{\sqrt{2n}} + \frac{1}{4n} + o(1/n).
$$

This equation shows that the average generalization loss is an increasing function of n. In this model, the distance from the true distribution to each model is constant, hence, if n is small, the posterior distribution is spread over all parameters. However, if n is large, then it concentrates on some parameter region by the random fluctuation. Such a phenomenon is called spontaneous symmetry breaking.

Remark 20. In this book, we mainly study the case when the log density ratio function has a relatively finite variance and show that the free energy and several losses are subject to the universal law. It may seem that Example 19 is a special or pathological exception, but, complicated and hierarchical statistical models used in deep learning may reveal the same phenomenon. For example, in some regression problem if a true distribution of Z

$$ Z = \exp(-X^2 - Y^2) + \mathcal{N}(0,1) $$

is statistically estimated by a neural network

$$ Z = \sum_{h=1}^{H} a_h \sigma(b_h X + c_h Y + d_h) + \mathcal{N}(0,1), $$

where σ is a sigmoidal function and $\{a_h, b_h, c_h, d_h\}$ is a parameter, then the true distribution is unrealizable by and singular for a statistical model. In this case, the same phenomenon shown by Example 19 occurs. Spontaneous symmetry breaking will be an important theme in Bayesian statistics in the future.

3.5   Problems

1. Let $x \in \mathbb{R}$ and $\{e_k(x); k = 1, 2, ...\}$ be a set of functions which satisfy

$$ \int e_k(x) e_\ell(x) q(x) dx = \delta_{k\ell}, $$

where if $k = \ell$ then $\delta_{k\ell} = 1$, and if $k \neq \ell$ then $\delta_{k\ell} = 0$. Let X be a random variable which is subject to a probability density $q(x)$ and $\{a_k \neq 0\}$ be a set of nonzero real values which satisfy

$$ \sum_{k=1}^{\infty} (a_k)^2 = 1. $$

Assume that a conditional density $q(y|x)$ and a statistical model $p(y|x, w)$ are respectively defined by

$$Y = \sum_{k=1}^{\infty} a_k e_k(X) + \mathcal{N}(0, 1^2),$$

$$Y = \sum_{k=1}^{K} w_k e_k(X) + \mathcal{N}(0, 1^2).$$

Then prove that $q(x)q(y|x)$ is unrealizable by and regular for $q(x)p(y|x, w)$.

2. Let $\beta > 0$ be a positive constant. For a given statistical model $p(x|w)$, a prior $\varphi(w)$, and a set of random variables X^n, a generalized posterior distribution of the inverse temperature β is defined by

$$p(w|X^n) = \frac{1}{Z_n(\beta)} \varphi(w) \prod_{i=1}^{n} p(X_i|w)^{\beta},$$

where

$$Z_n(\beta) = \int \varphi(w) \prod_{i=1}^{n} p(X_i|w)^{\beta} dw.$$

Let $\mathbb{E}_w^{\beta}[\ ]$ be the averaged value over $p^{(\beta)}(w|X^n)$. Then the generalized predictive density is defined by

$$p^{(\beta)}(x|X^n) = \mathbb{E}_w^{\beta}[p(x|w)].$$

Therefore, $\beta = 1$ results in the ordinary Bayesian estimation. The generalization loss, the training loss, the cross validation loss, and WAIC are respectively generalized by

$$G_n^{\beta} = -\mathbb{E}_X[\log p^{(\beta)}(X|X^n)],$$

$$T_n^{\beta} = -\frac{1}{n} \sum_{i=1}^{n} \log p^{(\beta)}(X_i|X^n),$$

$$C_n^{\beta} = -\frac{1}{n} \sum_{i=1}^{n} \log \frac{\mathbb{E}_w^{\beta}[p(X_i|w)^{1-\beta}]}{\mathbb{E}_w^{\beta}[p(X_i|w)^{-\beta}]},$$

$$W_n^{\beta} = T_n^{\beta} + \frac{\beta}{n} \sum_{i=1}^{n} \mathbb{V}_w^{\beta}[\log p(x|w)],$$

where $\mathbb{V}_w^\beta[\ \ ]$ is the variance over $p^{(\beta)}(w|X^n)$. The cumulant generating functions are also generalized by

$$G_n^\beta(\alpha) = \mathbb{E}_X[\log \mathbb{E}_w^\beta[p(X|w)^\alpha]], \tag{3.30}$$

$$T_n^\beta(\alpha) = \frac{1}{n}\sum_{i=1}^n \log \mathbb{E}_w^\beta[p(X_i|w)^\alpha]. \tag{3.31}$$

Assume that eqs.(3.24) and (3.25) hold for $G_n^\beta(\alpha)$ and $T_n^\beta(\alpha)$ instead of $G_n(\alpha)$ and $T_n(\alpha)$. Then prove the following equations.

$$G_n^\beta = -(G_n^\beta)'(0) - \frac{1}{2}(G_n^\beta)''(0) + o_p(\frac{1}{n}), \tag{3.32}$$

$$T_n^\beta = -(T_n^\beta)'(0) - \frac{1}{2}(T_n^\beta)''(0) + o_p(\frac{1}{n}), \tag{3.33}$$

$$C_n^\beta = -(T_n^\beta)'(0) + \frac{2\beta-1}{2}(T_n^\beta)''(0) + o_p(\frac{1}{n}), \tag{3.34}$$

$$W_n^\beta = -(T_n^\beta)'(0) + \frac{2\beta-1}{2}(T_n^\beta)''(0) + o_p(\frac{1}{n}). \tag{3.35}$$

These equations show that the universal law of the Bayesian statistics holds even if the posterior distribution is given by the inverse temperature $\beta > 0$.

3. Let us study the multinomial distribution and its prior defined by eq.(2.13) and eq.(2.14). Then by using.(2.18), it follows that

$$\log \mathbb{E}_w[p(x|w)^\alpha] = \log \int p(x|w)^\alpha p(w|X^n)dw$$

$$= \sum_{j=1}^N \log\Gamma(\alpha x^{(j)} + n_j + a_j) - \log\Gamma(n + \alpha + \sum_{j=1}^N a_j) + c_1,$$

where c_1 is a constant function of α. Assume that a true distribution is given by $p(x|w_0)$. Prove that the kth $(k \geq 1)$ cumulants are given by

$$G_n^{(k)}(0) = \sum_{j=1}^N w_{0j}\psi^{(k-1)}(n_j + a_j) - \psi^{(k-1)}(n + \sum_{j=1}^N a_j),$$

$$T_n^{(k)}(0) = \sum_{j=1}^N (n_j/n)\psi^{(k-1)}(n_j + a_j) - \psi^{(k-1)}(n + \sum_{j=1}^N a_j),$$

where $\psi^{(k-1)}(x)$ is the $(k-1)$th derivative of $\psi(x) = (\log\Gamma(x))'$. Then by using the asymptotic expansion

$$\psi^{(k-1)}(x) = O(1/x^{k-1}), \quad (k \geq 2),$$

prove that if $k \geq 2$,

$$\mathcal{G}_n^{(k)}(0) = O_p(1/n^{k-1}), \quad \mathcal{T}_n^{(k)}(0) = O_p(1/n^{k-1}).$$

Chapter 4

Regular Posterior Distribution

In this chapter, we study a special case when a true distribution $q(x)$ is regular for a statistical model $p(x|w)$ and the sample size n is large enough to ensure

$$\text{Posterior Distribution} \approx \text{Normal Distribution}$$

holds in the neighborhood of the optimal parameter w_0. In such a case, the asymptotic behaviors of the free energy or the minus log marginal likelihood are derived, and the generalization loss, the training losses, the cross validation loss, and WAIC are clarified.

(1) At first, we explain that the posterior distribution is divided into the essential and nonessential parts.

(2) Asymptotic expansion of the free energy is shown.

(3) Asymptotic expansions of the generalization loss, the training loss, the cross validation loss, and WAIC are proved.

(4) The mathematical proof is given for a basic Bayesian treatment.

(5) Point estimators such as the maximum likelihood or *a priori* are introduced.

A statistician who knows the conventional asymptotic theory can skip this section.

4.1   Division of Partition Function

In Bayesian theory, the parameter set is divided into the essential part and the nonessential part. The essential one is the set of the neighborhood of the optimal parameter, whereas the nonessential one is its complement.

99

Let $(q(x), p(x|w), \varphi(w))$ be a triple of a true probability density, a statistical model, and a prior. The average log loss function is defined by

$$L(w) = -\int q(x) \log p(x|w) dx.$$

We use a notation, $\nabla = (\partial/\partial w)$, then $\nabla L(w_0)$ and $\nabla^2 L(w_0)$ are a d-dimensional vector and a $d \times d$ matrix, respectively. The matrix $\nabla^2 L(w_0)$ is sometimes referred to as Hesse matrix of $L(w_0)$ at w_0. In this section, it is assumed that the set of all parameters W is a compact subset of $\mathbb{R}^d$ and that there exist both a unique parameter w_0 and an open subset U such that $w_0 \in U \subset W \subset \mathbb{R}^d$ and that w_0 minimizes $L(w)$. It is not assumed that the true density is realizable by a statistical model, in general. The probability density function of the optimal parameter is denoted by

$$p_0(x) = p(x|w_0).$$

Therefore $q(x) = p(x|w_0)$ does not hold in general. Also it is assumed that $\nabla \varphi(w)$ is a continuous function and that $\varphi(w_0) > 0$. The log density ratio function is

$$f(x, w) = \log \frac{p_0(x)}{p(x|w)} = \log \frac{p(x|w_0)}{p(x|w)}.$$

Hence $f(x, w_0) \equiv 0$. By using the average log density ratio function $K(w) = \mathbb{E}_X[f(X, w)]$,

$$L(w) = K(w) + L(w_0).$$

Therefore $K(w) \geq 0$ and $K(w)$ takes the minimum value zero if and only if $w = w_0$. The log likelihood ratio function is defined by

$$K_n(w) = \frac{1}{n} \sum_{i=1}^{n} f(X_i, w),$$

which satisfies $K_n(w_0) = 0$. For simple proof, we assume that $f(x, w)$ has a relatively finite variance and is a C^ℓ class function for sufficiently large ℓ, that is to say,

$$\nabla^\ell f(x, w)$$

is a continuous function of w. Further, it is assumed that, for a sufficiently large k,

$$\mathbb{E}_X[\sup_{w \in W} \|\nabla^\ell f(X, w)\|^k] < \infty. \tag{4.1}$$

Then $K(w)$ is also a C^ℓ class function. The main assumption of regularity is that

$$J \equiv \nabla^2 K(w_0)$$

is a positive definite matrix. By the assumption that $f(x, w)$ has a relatively finite variance, there exists $c_0 > 0$ such that

$$\mathbb{E}_X[f(X, w)^2] \leq c_0 K(w),$$

resulting that a function

$$a(x, w) = \frac{K(w) - f(X_i, w)}{\sqrt{K(w)}}$$

is well defined for $w \neq w_0$ and

$$\sup_{w \neq w_0} \mathbb{E}_X\left[a(X, w)^2\right] < \infty. \tag{4.2}$$

Remark 21. In the neighborhood of w_0,

$$K(w) = \frac{1}{2}(w - w_0)J(w - w_0) + o(\|w - w_0\|^3).$$

By the condition that $f(x, w)$ has a relatively finite variance, there exists $c_1 > 0$ such that

$$\mathbb{E}_X[f(X, w)^2] \leq c_1 \|w - w_0\|^2.$$

Hence eq.(4.2) holds. The function $a(x, w)$ is bounded but may be discontinuous at $w = w_0$. However, it can be made well-defined as a function of the generalized polar coordinate $(w - w_0) = r\Theta$, where $r = \|w - w_0\|$ and $\Theta = (w - w_0)/r$.

Example 20. In order to illustrate functions defined in the foregoing statement, we study a simple case,

$$p(x, y|u, v) = \frac{1}{2\pi} \exp\left(-\frac{1}{2}((x - u)^2 + (y - v)^2)\right)$$

and $q(x, y) = p(x, y|0, 0)$. Then $w_0 = (0, 0)$. The log density ratio function, and its average function are respectively given by

$$
\begin{aligned}
f(x, y, u, v) &= \frac{1}{2}(u^2 + v^2 - 2ux - 2vy), \\
K(u, v) &= \frac{1}{2}(u^2 + v^2).
\end{aligned}
$$

The log likelihood ratio function is

$$K_n(u,v) = \frac{1}{2}(u^2 + v^2) - u\left(\frac{1}{\sqrt{n}}\sum_{i=1}^{n} X_i\right) - v\left(\frac{1}{\sqrt{n}}\sum_{i=1}^{n} Y_i\right).$$

The function $a(x, y, a, b)$ is

$$a(x, y, u, v) = \frac{\sqrt{2}\cdot(ux + vy)}{\sqrt{u^2 + v^2}}.$$

For a given $(x, y) \neq (0, 0)$, $a(x, y, u, v)$ is a bounded but discontinuous function of (u, v) in the neighborhood of $(u, v) = (0, 0)$. However, by using

$$u = r\cos\theta, \quad v = r\sin\theta,$$

it follows that

$$a(x, y, r, \theta) = \sqrt{2}\,(x\cos\theta + y\sin\theta)$$

is a well-defined function. Note that $(u, v) = (0, 0)$ corresponds to $r = 0$ and θ = free. If a true density is regular for a statistical model, then the same transform is always employed. In the following chapters, we show that, even if a true density is not regular for a statistical model, the generalized procedure called resolution of singularities made by algebraic geometry can be applied.

We define a function

$$\gamma_n(w) = \frac{1}{\sqrt{n}}\sum_{i=1}^{n} a(X_i, w),$$

and assume that it satisfies the asymptotic expectation condition with index k. For the asymptotic expectation condition, see Section 10.5. That is to say, we assume that

$$\gamma_n = \sup_{w \neq w_0} |\gamma_n(w)|$$

satisfies $\mathbb{E}[(\gamma_n)^{k+\varepsilon_0}] < \infty$ for some $\varepsilon_0 > 0$. Firstly, we analyze the normalized partition function defined by

$$Z_n^{(0)} = \int \exp(-nK_n(w))\varphi(w)dw.$$

For a given positive value $\epsilon > 0$, we define

$$W_1 = \{w \in W; \|w - w_0\| < \epsilon\},$$
$$W_2 = \{w \in W; \|w - w_0\| \geq \epsilon\}.$$

Then by defining

$$Z_n^{(1)} = \int_{W_1} \exp(-nK_n(w))\varphi(w)dw,$$

$$Z_n^{(2)} = \int_{W_2} \exp(-nK_n(w))\varphi(w)dw,$$

the normalized partition function is equal to

$$Z_n^{(0)} = Z_n^{(1)} + Z_n^{(2)}. \tag{4.3}$$

Here $Z_n^{(1)}$ and $Z_n^{(2)}$ are the integrations of parameters in a neighborhood of the optimal parameter w_0 and in its complement respectively.

In regular theory, the posterior distribution becomes to be accumulated on a neighborhood of the optimal parameters when $n \to \infty$. In this chapter, we define a positive real value ϵ as a function of n,

$$\epsilon = \frac{1}{n^{2/5}}.$$

Then

$$\lim_{n \to \infty} \epsilon = 0,$$
$$\lim_{n \to \infty} \sqrt{n}\epsilon = \infty.$$

In the following, we show that, when $n \to \infty$,

$$Z_n^{(1)} >> Z_n^{(2)},$$

hence $Z_n^{(1)}$ and $Z_n^{(2)}$ are called the essential and nonessential parts of the normalized partition function respectively.

Firstly we study the nonessential part $Z_n^{(2)}$.

Lemma 9. *Let $J_1 > 0$ be the minimum eigenvalue of the matrix $J = \nabla^2 K(w_0)$ and K_1 be the maximum value of $K(w)$ in W. If n is sufficiently large,*

$$Z_n^{(2)} \leq \exp(-(J_1/4)n^{1/5} + \gamma_n^2/2),$$
$$Z_n^{(2)} \geq \exp(-2K_1 n - \gamma_n^2/2).$$

Hence for arbitrary $k \geq 0$, the convergence in probability $n^k Z_n^{(2)} \to 0$ holds.

Proof. By the definition of $\gamma_n(w)$,

$$nK_n(w) = nK(w) - \sqrt{nK(w)}\,\gamma_n(w).$$

By using

$$\sqrt{nK(w)}\,\gamma_n(w) \le \frac{1}{2}(nK(w) + \gamma_n(w)^2),$$

it follows that

$$\begin{aligned}
nK_n(w) &\ge& nK(w)/2 - \gamma_n^2/2, \\
nK_n(w) &\le& 3nK(w)/2 + \gamma_n^2/2.
\end{aligned}$$

In the neighborhood of the origin,

$$nK(w) = \frac{n}{2}\|J^{1/2}(w - w_0)\|^2 + O(n\|w - w_0\|^3).$$

Hence, in $\|w - w_0\| > n^{-2/5}$, for sufficiently large n,

$$(J_1/4)n^{1/5} \le nK(w) \le nK_1,$$

and

$$\exp(-nK_1/2) \le \int_{W_2} \varphi(w)dw.$$

Then by the definition,

$$Z_n^{(2)} = \int_{\|w-w_0\|\ge n^{-2/5}} \exp(-nK_n(w))\varphi(w)dw,$$

Lemma 9 is obtained.   □

Secondly, we study the essential part of the normalized partition function which is the integration on the set, $\|w - w_0\| < n^{-2/5}$. An empirical process $\eta_n(w)$ is defined by

$$\eta_n(w) = \frac{1}{\sqrt{n}}\sum_{i=1}^{n}\{K(w) - f(X_i, w)\}. \qquad (4.4)$$

Then

$$nK_n(w) = nK(w) - \sqrt{n}\eta_n(w).$$

We assume that $\nabla \eta_n(w)$ and $\nabla^2 \eta_n(w)$ satisfy the asymptotic expectation condition with sufficiently large index k. Two random variables are defined by

$$\eta_n^{(2)} = \frac{1}{2} \sup_{w \in W_1} \|\nabla^2 \eta_n(w)\|,$$

$$\xi_n = J^{-1/2} \nabla \eta_n(w_0).$$

Then $\mathbb{E}[\eta_n^{(2)}] < \infty$ and $\xi_n \to \xi$ in distribution hold where ξ is the random variable that is subject to the normal distribution. Also we define a function $\delta_n(w)$ by

$$\delta_n(w) = n\{K(w) - \frac{1}{2}\|J^{1/2}(w - w_0)\|^2\}$$
$$+ \sqrt{n}\{\eta_n(w) - \nabla \eta_n(w_0) \cdot (w - w_0)\}. \tag{4.5}$$

By the definition, it follows that

$$nK_n(w) = \frac{n}{2}\|J^{1/2}(w - w_0)\|^2 - \sqrt{n}\nabla \eta_n(w_0) \cdot (w - w_0) + \delta_n(w).$$

By the assumption that $\nabla^3 K(w)$ is a continuous function,

$$k^{(3)} \equiv \frac{1}{6} \sup_{w \in W_1} \|\nabla^3 K(w)\| < \infty.$$

Lemma 10. *The following inequalities hold.*

$$\sup_{w \in W_1} |\delta_n(w)| \leq n^{-1/5}k^{(3)} + n^{-3/5}\eta_n^{(2)}, \tag{4.6}$$

$$\sup_{w \in W_1} \|\nabla \delta_n(w)\| \leq 3n^{1/5}k^{(3)} + 2n^{-1/5}\eta_n^{(2)}. \tag{4.7}$$

Proof. By applying the mean value theorem to $K(w)$ and $\eta_n(w)$, there exist $w^*, w^{**} \in W_1$ such that

$$K(w) = \frac{1}{2}\|J^{1/2}(w - w_0)\|^2 + \frac{1}{6}\nabla^3 K(w^*)(w - w_0)^3,$$

$$\eta_n(w) = \nabla \eta_n(w_0) \cdot (w - w_0) + \frac{1}{2}\nabla^2 \eta_n(w^{**})(w - w_0)^2,$$

hence eq.(4.6) is derived. Also by applying the mean value theorem to $\nabla K(w)$ and $\nabla \eta_n(w)$, there exist $w', w'' \in W_1$ such that

$$\nabla K(w) = J(w - w_0) + \frac{1}{2}\nabla^3 K(w')(w - w_0)^2,$$

$$\nabla \eta_n(w) = \nabla \eta_n(w_0) + \nabla^2 \eta_n(w'')(w - w_0),$$

hence eq.(4.7) is derived. $\square$

We assume that a prior $\varphi(w)$ is a C^1 class function and that $\varphi(w_0) > 0$. Then

$$\varphi^{(1)} \equiv \sup_{w \in W_1} \|\nabla \varphi(w)\| < \infty.$$

The following lemma shows the asymptotic behavior of the essential part.

Lemma 11. *The essential part of the normalized partition function satisfies inequalities,*

$$Z_n^{(1)} \leq \frac{(2\pi)^{d/2}}{n^{d/2}(\det J)^{1/2}}\left(\varphi(w_0) + \frac{\varphi^{(1)}}{n^{2/5}}\right)$$

$$\times \exp\left(\frac{1}{2}\|\xi_n\|^2 + \frac{k^{(3)}}{n^{1/5}} + \frac{\eta_n^{(2)}}{n^{3/5}}\right),$$

$$Z_n^{(1)} \geq \frac{(2\pi)^{d/2}}{n^{d/2}(\det J)^{1/2}}\left(\varphi(w_0) - \frac{\varphi^{(1)}}{n^{2/5}}\right)$$

$$\times \exp\left(\frac{1}{2}\|\xi_n\|^2 - \frac{k^{(3)}}{n^{1/5}} - \frac{\eta_n^{(2)}}{n^{3/5}}\right) \Phi_n(\xi_n),$$

where

$$\Phi_n(\xi_n) \equiv \frac{1}{(2\pi)^{d/2}} \int_{\|J^{1/2}w\| < n^{3/5}} \exp(-\|w - \xi_n\|^2/2)dw, \qquad (4.8)$$

which satsifies

$$\Phi_n(\xi_n) \geq \frac{\exp(-\|\xi_n\|^2)}{2(2\pi)^{d/2}}$$

for sufficiently large n. Therefore the convergence in probability holds,

$$n^{d/2} Z_n^{(1)} \exp(-\|\xi_n\|^2/2) \to \frac{(2\pi)^{d/2}}{(\det J)^{1/2}} \varphi(w_0).$$

Proof. In the region $W_1 = \{w \in W; \|w - w_0\| < n^{-2/5}\}$, by the mean value theorem,

$$\varphi(w) \leq \varphi(w_0) + n^{-2/5}\varphi^{(1)},$$
$$\varphi(w) \geq \varphi(w_0) - n^{-2/5}\varphi^{(1)}.$$

Also in the region W_1,

$$nK_n(w) \leq \frac{1}{2}\|(nJ)^{1/2}(w - w_0) - \xi_n\|^2 - \frac{\|\xi_n\|^2}{2} + \frac{k^{(3)}}{n^{1/5}} + \frac{\eta_n^{(2)}}{n^{3/5}},$$

$$nK_n(w) \geq \frac{1}{2}\|(nJ)^{1/2}(w - w_0) - \xi_n\|^2 - \frac{\|\xi_n\|^2}{2} - \frac{k^{(3)}}{n^{1/5}} - \frac{\eta_n^{(2)}}{n^{3/5}}.$$

By putting $w' = (nJ)^{1/2}(w - w_0)$,

$$\int_{w \in W_1} \exp(-\frac{1}{2}\|(nJ)^{1/2}(w - w_0) - \xi_n\|^2)dw = \frac{(2\pi)^{d/2}\Phi_n(\xi_n)}{n^{d/2}(\det J)^{1/2}}.$$

Since $\Phi_n(\ )$ is the probability of the set $\|J^{-1/2}w'\| < n^{3/5}$, $\Phi_n(\xi_n) < 1$. Moreover,

$$\Phi_n(\xi_n) \geq \frac{\exp(-\|\xi_n\|^2)}{(2\pi)^{d/2}} \int_{\|J^{1/2}w'\| < n^{3/5}} \exp(-\|w'\|^2)dw'$$

which completes Lemma. □

In this section we proved that the essential and nonessential normalized partition functions satisfy

$$\begin{aligned} Z_n^{(1)} &\propto 1/n^{d/2}, \\ Z_n^{(2)} &\leq \exp(-n^{1/5}), \end{aligned}$$

as random variables.

4.2 Asymptotic Free Energy

In this section, we derive the asymptotic expansion of the free energy or the minus log marginal likelihood F_n. A matrix I is defined by

$$I = \mathbb{E}_X[\nabla f(x, w_0)(\nabla f(x, w_0))^T].$$

Then by the definition,

$$\mathbb{E}[\frac{1}{n} \sum_{i=1}^{n} \nabla f(X_i, w_0)(\nabla f(X_i, w_0))^T] = I,$$

and by the central limit theorem,

$$\frac{1}{n} \sum_{i=1}^{n} \nabla f(X_i, w_0)(\nabla f(X_i, w_0))^T = I + O_p(1/n^{3/2}).$$

If $q(x)$ is realizable by $p(x|w)$, then I is called the Fisher information matrix at w_0 and $I = J$. However, in general, $I \neq J$. By the definition of I, it is positive semidefinite. In the regularity condition of Bayesian theory, the eigenvalue of J positive, but I may contain a zero eigenvalue.

Theorem 4. *If the regularity condition holds, then the Bayesian free energy F_n satisfies*

$$
\begin{aligned}
F_n &= nL_n(w_0) + \frac{d}{2}\log n - \log\varphi(w_0) - \frac{d}{2}\log(2\pi) \\
&\quad + \frac{1}{2}\log\det J - \frac{1}{2}\|\xi_n\|^2 + o_p(1). \quad\quad (4.9)
\end{aligned}
$$

$$
\begin{aligned}
\mathbb{E}[F_n] &= nL(w_0) + \frac{d}{2}\log n - \log\varphi(w_0) - \frac{d}{2}\log(2\pi) \\
&\quad + \frac{1}{2}\log\det J - \frac{1}{2}\mathrm{tr}(J^{-1}I) + o(1). \quad\quad (4.10)
\end{aligned}
$$

Proof. By the assumption, ξ_n and $\nabla^2\eta_n(w)$ converge in distribution and their expectation values also converge. By the definition,

$$
F_n = F_n^{(0)} + nL_n(w_0),
$$

where $F_n^{(0)}$ is a normalized free energy, which is given by the essential and nonessential parts,

$$
\begin{aligned}
F_n^{(0)} &= -\log Z_n^{(0)} \\
&= -\log(Z_n^{(1)} + Z_n^{(2)}).
\end{aligned}
$$

When $n \to \infty$, convergence in probability $\Phi_n(\xi_n) \to 1$ holds, where $\Phi_n(\xi_n)$ is defined by eq.(4.8). By Lemma 9 and 11, covergences in probability

$$
\begin{aligned}
n^{d/2}Z_n^{(1)}\exp(-\frac{1}{2}\|\xi_n\|^2) &\to \frac{(2\pi)^{d/2}}{(\det J)^{1/2}}\varphi(w_0), \\
n^{d/2}Z_n^{(2)} &\to 0,
\end{aligned}
$$

hold. It follows that eq.(4.9) holds. Let us prove eq.(4.10). Let ξ be a random variable which is subject to the normal distribution whose average and covariance are equal to those of ξ_n. Then by the central limit theorem, $\xi_n \to \xi$ in distribution holds. By the convergence in distribution,

$$
n^{d/2}Z_n^{(1)} \to \frac{(2\pi)^{d/2}}{(\det J)^{1/2}}\varphi(w_0)\exp(\frac{1}{2}\|\xi\|^2),
$$

and convergence in probability $n^{d/2}Z_n^{(2)} \to 0$, the convergence in distribution

$$
n^{d/2}Z_n^{(0)} \to \frac{(2\pi)^{d/2}}{(\det J)^{1/2}}\varphi(w_0)\exp(\frac{1}{2}\|\xi\|^2)
$$

holds. Hence $F_n^{(0)} - (d/2)\log n$ also converges in distribution. In order to prove the convergence of the expected value $\mathbb{E}[(F_n^{(0)} - (d/2)\log n)]$, it is sufficient to prove that the sequence of random variables $(F_n^{(0)} - (d/2)\log n)$ is uniformly asymptotically integrable. To prove it is uniformly asymptotically intgerable, it is sufficient to prove $\mathbb{E}[|F_n^{(0)} - (d/2)\log n|^{1+\varepsilon}] < \infty$ for some $\varepsilon > 0$. (See Section 10.5). We define two random variables,

$$\begin{aligned} A_n &= \log(n^{d/2} Z_n^{(1)}), \\ B_n &= \log(n^{d/2} Z_n^{(2)}). \end{aligned}$$

Then

$$(d/2)\log n - F_n^{(0)} = \log(e^{A_n} + e^{B_n}).$$

For arbitrary real numbers x, y,

$$x \le \log(e^x + c^y) \le \max(x, y) + \log 2.$$

Hence

$$|\log(e^x + e^y)| \le \max(|x|, |\max(x, y)| + \log 2).$$

Therefore

$$|\log(e^{A_n} + e^{B_n})| \le \max(|A_n|, |\max(A_n, B_n)|) + \log 2.$$

By Lemma 9 and 11, and

$$\Phi_n(\xi_n) \ge \exp(-\|\xi_n\|^2) \int_{\|w-w_0\|<1} \exp(-\|w\|^2)dw,$$

there exists constants c_1, c_2, c_3 such that

$$\begin{aligned} A_n &\le c_1 + \|\xi_n\|^2/2 + \eta_n^{(2)}/n^{3/5}, \\ A_n &\ge c_2 - \|\xi_n\|^2/2 - \eta_n^{(2)}/n^{3/5}, \\ \max(A_n, B_n) &\le c_3 + \|\xi_n\|^2/2 + \eta_n^{(2)}/n^{3/5} + \gamma_n^2/2. \end{aligned}$$

Then by

$$\max(A_n, B_n) \ge A_n,$$

there exists $c_4 > 0$ such that

$$|(d/2)\log n - F_n^{(0)}| \le \|\xi_n\|^2/2 + \eta_n^{(2)}/n^{3/5} + c_4.$$

By the assumption ξ_n and $\eta_n^{(2)}$ have the asymptotic expectation condition with index k, $\mathbb{E}[|F_n^{(0)} - (d/2)\log.n|^{1+\epsilon}] < \infty$ for some $\epsilon > 0$. Lastly, since

$$\mathbb{E}[\nabla\eta_n(w_0)(\nabla\eta_n(w_0))^T] = \int (\nabla f(x, w_0))(\nabla f(x, w_0))^T q(x)dx,$$

it follows that

$$\begin{aligned}\mathbb{E}[\|\xi_n\|^2] &= \mathbb{E}[\mathrm{tr}(J^{-1}\nabla\eta_n(w_0)(\nabla\eta_n(w_0))^T)] \\ &= \mathrm{tr}(J^{-1}\mathbb{E}[\nabla\eta_n(w_0)(\nabla\eta_n(w_0))^T]),\end{aligned}$$

which completes the theorem. □

Remark 22. In this section, on the regularity condition, $J > 0$ and $\varphi(w_0) > 0$, we proved that

$$\begin{aligned}F_n &= nL_n(w_0) + \frac{d}{2}\log n - \log\varphi(w_0) - \frac{d}{2}\log(2\pi) \\ &\quad + \frac{1}{2}\log\det J - \frac{1}{2}\|\xi_n\|^2 + o_p(1)\end{aligned}$$

and

$$\mathbb{E}[\|\xi_n\|^2] = \mathrm{tr}(IJ^{-1}) + o(1).$$

If $\det J = 0$, $\det J = \infty$, $\varphi(w_0) = 0$, or $\varphi(w_0) = \infty$, then this asymptotic expansion does not hold because

$$\log\det J, \quad \log\varphi(w_0)$$

are not finite. For the case when the regularity condition is not satisfied, see the following sections.

Example 21. Let $x, w \in \mathbb{R}^M$ and $y \in \mathbb{R}$, and a statistical model is

$$p(y|x, w) = \frac{1}{(2\pi)^{1/2}} \exp(-\frac{1}{2}(y - w \cdot x)^2). \tag{4.11}$$

Assume that X is subject to some density $q(x)$. The matrix J of the statistical model eq.(4.11) is

$$J_{jk} = \int x_j x_k q(x)dx.$$

The support of $q(x)$ is defined by

$$\mathrm{supp}\, q \equiv \overline{\{x \in \mathbb{R}^M \;;\; q(x) > 0\}},$$

where $\overline{A}$ of a set $A \subset \mathbb{R}^M$ is the closure of A. If the support of $q(x)$ is contained in a subspace of $\mathbb{R}^M$ whose dimension is smaller than M, then $\det J = 0$. In real world problems which are defined on high dimensional space, the support of $q(x)$ is sometimes contained in a low dimensional subspace. In such cases, the regular asymptotic theory does not hold. However, such cases can be analyzed by the general theory.

Example 22. Let $p(x|w)$ be a statistical model and assume that the optimal parameter $w_0 = 0$. In statistical estimation, sometimes a prior

$$\varphi(w) \propto |w|^{\alpha - 1}$$

is employed, where $\alpha > 0$ is a hyperparameter. Then the regularity condition is satisfied if and only if $\alpha = 1$. The cases $\alpha \neq 1$ can be analyzed by the general theory.

4.3 Asymptotic Losses

In this section, we show asymptotic behaviors of the generalization, cross validation, and training losses when $n \to \infty$, based on the regularity condition. The following lemma is necessary in this section.

Lemma 12. *Assume the regularity condition. Let k ($k \geq 2$) be an integer and $g(x, w)$ be a function which satisfies $g(x, w_0) = 0$ and assume that, for a sufficiently large integer ℓ,*

$$\mathbb{E}_X[\sup_{w \in W} |g(X, w)|^\ell] < \infty,$$

$$\mathbb{E}_X[\sup_{w \in W} \|\nabla g(X, w)\|^\ell] < \infty.$$

We use a notation,

$$L(X, \alpha) \equiv \frac{\mathbb{E}_w[|g(X, w)|^k \exp(-\alpha f(X, w))]}{\mathbb{E}_w[\exp(-\alpha f(X, w))]}.$$

Then, there exists $\varepsilon > 0$ such that

$$\mathbb{E}\left[\left(n^{k/2} \sup_{|\alpha| \leq 1} \mathbb{E}_X[L(X, \alpha)]\right)^{1+\varepsilon}\right] < \infty \tag{4.12}$$

$$\mathbb{E}\left[\left(n^{k/2} \sup_{|\alpha| \leq 1} \frac{1}{n} \sum_{i=1}^{n} L(X_i, \alpha)\right)^{1+\varepsilon}\right] < \infty. \tag{4.13}$$

The proof of this lemma is given in the following section. In this section, we employ Lemma 12 and prove asymptotic expansions of the generalization, cross validation, and training losses.

Remark 23. The special case $\alpha = 0$ and $g(X, w) = g(w)$ shows that $\mathbb{E}_w[n^{k/2}|g(w)|^k]$ is asymptotically uniformly integrable. Therefore, if the random variable $\mathbb{E}_w[n^{k/2}g(w)^k]$ converges in distribution, then a sequence $\mathbb{E}\,\mathbb{E}_w[n^{k/2}g(w)^k]$ also converges.

Firstly, the basic Theorem 3 in the foregoing chapter is proved in regular cases.

Theorem 5. *Based on the regularity condition, the assumptions of the basic Theorem 3 are satisfied,*

$$\sup_{|\alpha|\leq 1}\left|\left(\frac{d}{d\alpha}\right)^k \mathcal{G}_n(\alpha)\right| \;\leq\; O_p(\frac{1}{n^{k/2}}), \tag{4.14}$$

$$\sup_{|\alpha|\leq 1}\left|\left(\frac{d}{d\alpha}\right)^k \mathcal{T}_n(\alpha)\right| \;\leq\; O_p(\frac{1}{n^{k/2}}), \tag{4.15}$$

$$\mathbb{E}[\sup_{|\alpha|\leq 1}\left|\left(\frac{d}{d\alpha}\right)^k \mathcal{G}_n(\alpha)\right|] \;\leq\; O(\frac{1}{n^{k/2}}), \tag{4.16}$$

$$\mathbb{E}[\sup_{|\alpha|\leq 1}\left|\left(\frac{d}{d\alpha}\right)^k \mathcal{T}_n(\alpha)\right|] \;\leq\; O(\frac{1}{n^{k/2}}). \tag{4.17}$$

Proof. By applying Lemma 8 and Lemma 12 to the case $g(x, w) = f(x, w)$, where $f(x, w)$ is the log density ratio function, eq.(4.16) and eq.(4.17) are immediately derived. By Lemma 12, the random variables

$$n^{k/2} \sup_{|\alpha|\leq 1} \mathbb{E}_X[L(X, \alpha)],$$

$$n^{k/2} \sup_{|\alpha|\leq 1} \frac{1}{n}\sum_{i=1}^n L(X_i, \alpha)$$

are asymptotically uniformly integrable, resulting that they are also uniformly tight. Hence eq.(4.14) and eq.(4.15) are obtained.   □

Definition 11. Let $(i_1, i_2, ..., i_t)$ be a set of integers and w_j be the jth element of the vector w. The constant $\Phi(i_1, i_2, ..., i_t)$ is defined by

$$\Phi(i_1, i_2, ..., i_t) = \frac{1}{(2\pi)^{d/2}} \int w_{i_1} w_{i_2} \cdots w_{i_t} \exp(-\frac{\|w\|^2}{2})dw,$$

which is the expected value of $w_{i_1} w_{i_2} \cdots w_{i_t}$ with respect to the normal distribution. In other words, if X is subject to the d dimensional normal distribution whose average and covariance matrix are zero and identity respectively, then

$$\Phi(i_1, i_2, ..., i_t) = \mathbb{E}_X[X_{i_1} X_{i_2} \cdots X_{i_t}].$$

By using this definition, in oder to derive the asymptotic behaviors of the generalization, cross validation, and training losses, we first show the correlations of parameters in the posterior distribution.

Lemma 13. *let w_0 be the optimal parameter and $\xi_n = J^{-1/2} \nabla \eta_n(w_0)$. For an arbitrary $i_1, i_2, ..., i_t$, a function $\Sigma_n(w)$ is defined by*

$$\Sigma_n(w) = \{((nJ)^{1/2}(w - w_0) - \xi_n)_{i_1}\} \cdots \{((nJ)^{1/2}(w - w_0) - \xi_n)_{i_t}\}.$$

Then their averages by the posterior distribution converge in probability,

$$\mathbb{E}_w[\Sigma_n(w)] \to \Phi(i_1, i_2, ..., i_t).$$

Proof. For the essential and nonessential sets of parameters W_1 and W_2, we define

$$Z_n^{(1)}(i_1, ..., i_t) = \int_{W_1} \Sigma_n(w) \exp(-nK_n(w)) \varphi(w) dw,$$

$$Z_n^{(2)}(i_1, ..., i_t) = \int_{W_2} \Sigma_n(w) \exp(-nK_n(w)) \varphi(w) dw.$$

By the same way as Lemma 9 and 11, the convergences in probability can be derived,

$$n^{d/2} Z_n^{(1)}(i_1, ..., i_t) \exp(-\frac{1}{2}\|\xi_n\|^2) \to \frac{(2\pi)^{d/2}}{(\det J)^{1/2}} \Phi(i_1, i_2, ..., i_t) \varphi(w_0),$$

$$n^{d/2} Z_n^{(2)}(i_1, ..., i_t) \exp(-\frac{1}{2}\|\xi_n\|^2) \to 0.$$

Note that for the case $(i_1, i_2, ..., i_t)$ is the empty set, we define $Z_n^{(1)} = Z_n^{(1)}(\varnothing)$ and $Z_n^{(2)} = Z_n^{(2)}(\varnothing)$. Then

$$\mathbb{E}_w[\Sigma_n(w)] = \frac{Z_n^{(1)}(i_1, ..., i_t) + Z_n^{(2)}(i_1, ..., i_t)}{Z_n^{(1)} + Z_n^{(2)}},$$

$$\to \Phi(i_1, i_2, ..., i_t),$$

which completes the lemma. $\square$

Lemma 14. *The following hold.*

$$\mathbb{E}_w[(w - w_0)] = (nJ)^{-1/2}\xi_n + o_p(1/\sqrt{n}),$$

$$\mathbb{E}_w[(w - w_0)(w - w_0)^T] = \frac{1}{n}(J^{-1} + J^{-1/2}\xi_n\xi_n^T J^{-1/2}) + o_p(1/n),$$

$$\mathbb{E}\,\mathbb{E}_w[(w - w_0)(w - w_0)^T] = \frac{1}{n}(J^{-1} + J^{-1/2}IJ^{-1/2}) + o(1/n).$$

Proof. By Lemma 13, if $t = 1$ then $\Phi(i_1) = 0$, hence

$$(nJ)^{1/2}\mathbb{E}_w[(w - w_0)] - \xi_n \to 0, \tag{4.18}$$

which shows the first equation. By Lemma 13, if $t = 2$ and $\Phi(i, j) = \delta_{ij}$, the convergence in probability holds,

$$\mathbb{E}_w[\{((nJ)^{1/2}(w - w_0) - \xi_n)\}\{((nJ)^{1/2}(w - w_0) - \xi_n)\}^T] \to I_d,$$

where I_d is the $d \times d$ identity matrix. By eq.(4.18) and convergence in distribution of ξ_n,

$$nJ^{1/2}\mathbb{E}_w[(w - w_0)(w - w_0)^T]J^{1/2} - \xi_n\xi_n^T \to I_d.$$

By Lemma 12,

$$nJ^{1/2}\mathbb{E}\,\mathbb{E}_w[(w - w_0)(w - w_0)^T]J^{1/2} - I \to I_d,$$

which completes the lemma. $\square$

Lemma 15. *The cumulants satisfy the relations,*

$$\mathcal{G}_n'(0) = -L(w_0) - \frac{d + \|\xi_n\|^2}{2n} + o_p(1/n), \tag{4.19}$$

$$\mathcal{G}_n''(0) = \frac{\text{tr}(IJ^{-1})}{n} + o_p(1/n), \tag{4.20}$$

$$\mathcal{T}_n'(0) = -L_n(w_0) + \frac{-d + \|\xi_n\|^2}{2n} + o_p(1/n), \tag{4.21}$$

$$\mathcal{T}_n''(0) = \frac{\text{tr}(IJ^{-1})}{n} + o_p(1/n). \tag{4.22}$$

Proof. By the mean value theorem and Lemma 14,

$$\mathbb{E}_w[K(w)] = \mathbb{E}_w[\frac{1}{2}\|J^{1/2}(w - w_0)\|^2] + o_p(1/n)$$

$$= \frac{1}{2}\mathbb{E}_w[\text{tr}(J(w - w_0)(w - w_0)^T)] + o_p(1/n)$$

$$= \frac{1}{2n}(d + \|\xi_n\|^2) + o_p(1/n).$$

Also by the mean value theorem,

$$
\begin{aligned}
f(x,w) &= (w - w_0) \cdot \nabla f(x, w_0) \\
&\quad + \frac{1}{2}\mathrm{tr}\Big(\nabla^2 f(x, w_0)(w - w_0)(w - w_0)^T\Big) \\
&\quad + O(\|w - w_0\|^3).
\end{aligned}
$$

Hence,

$$
\begin{aligned}
\mathbb{E}_w[f(x,w)] &= \mathbb{E}_w[(w - w_0)] \cdot \nabla f(x, w_0) \\
&\quad + \frac{1}{2}\mathrm{tr}\Big(\nabla^2 f(x, w_0)\mathbb{E}_w[(w - w_0)(w - w_0)^T]\Big) \\
&\quad + o_p(1/n) \\
&= \frac{1}{\sqrt{n}} J^{-1/2}\xi_n \cdot \nabla f(x, w_0) \\
&\quad + \frac{1}{2n}\mathrm{tr}\Big(\nabla^2 f(x, w_0)(J^{-1} + J^{-1/2}\xi_n\xi_n^T J^{-1/2})\Big) \\
&\quad + o(1/n)
\end{aligned}
$$

and

$$
\begin{aligned}
\mathbb{E}_w[f(x,w)^2] &= \mathbb{E}_w[((w - w_0) \cdot \nabla f(X_i, w_0))^2] + o_p(1/n) \\
&= \mathrm{tr}\Big(\mathbb{E}_w[(w - w_0)(w - w_0)^T]\nabla f(x, w_0)(\nabla f(x, w_0))^T\Big) \\
&\quad + o_p(1/n) \\
&= \frac{1}{n}\mathrm{tr}\Big((J^{-1} + J^{-1/2}\xi_n\xi_n^T J^{-1/2})\nabla f(x, w_0)\nabla f(x, w)^T\Big) \\
&\quad + o_p(1/n).
\end{aligned}
$$

Then by using

$$
\frac{1}{\sqrt{n}}\sum_{i=1}^{n} \nabla f(X_i, w_0) = -J^{1/2}\xi_n, \tag{4.23}
$$

$$
\frac{1}{n}\sum_{i=1}^{n} \nabla^2 f(X_i, w_0) = J + o_p(1) \tag{4.24}
$$

$$
\mathbb{E}_X[\nabla^2 f(X, w_0)] = J. \tag{4.25}
$$

It follows that

$$
\frac{1}{n}\sum_{i=1}^{n}\mathbb{E}_w[f(X_i,w)] = -\frac{\|\xi_n\|^2}{n} + \frac{\mathrm{tr}(J(J^{-1}+J^{-1/2}\xi_n\xi_n^T J^{-1/2}))}{2n}
$$
$$
+o_p(1/n)
$$
$$
= \frac{d-\|\xi_n\|^2}{2n} + o_p(1/n).
$$

By using the results above,

$$
\frac{1}{n}\sum_{i=1}^{n}\mathbb{E}_w[f(X_i,w)]^2 = \frac{1}{n^2}\sum_{i=1}^{n}(J^{-1/2}\xi_n \cdot \nabla f(X_i,w_0))^2 + o_p(1/n)
$$
$$
= \frac{1}{n^2}\sum_{i=1}^{n}\mathrm{tr}\Big((J^{-1/2}\xi_n\xi_n^T J^{-1/2})\nabla f(X_i,w_0)\nabla f(X_i,w_0)^T\Big) + o_p(1/n)
$$
$$
= \frac{1}{n}\mathrm{tr}\Big((J^{-1/2}\xi_n\xi_n^T J^{-1/2})I\Big) + o_p(1/n)
$$

and

$$
\frac{1}{n}\sum_{i=1}^{n}\mathbb{E}_w[f(X_i,w)^2]
$$
$$
= \frac{1}{n^2}\sum_{i=1}^{n}\mathrm{tr}\Big((J^{-1}+J^{-1/2}\xi_n\xi_n^T J^{-1/2})\nabla f(X_i,w_0)\nabla f(X_i,w_0)^T\Big)
$$
$$
+o_p(1/n)
$$
$$
= \frac{1}{n}\mathrm{tr}\Big((J^{-1}+J^{-1/2}\xi_n\xi_n^T J^{-1/2})I\Big) + o_p(1/n)
$$

Then by using eq.(3.14) through eq.(3.17), the lemma is derived.   □

Theorem 6. *(Regular asymptotic theory) Assume that a true distribution is regular for a statistical model. Then the generalization loss, the cross validation loss, the training loss, and WAIC are asymptotically equal to*

$$
G_n = L(w_0) + \frac{d+\|\xi_n\|^2-\mathrm{tr}(IJ^{-1})}{2n} + o_p(1/n), \tag{4.26}
$$

$$
T_n = L_n(w_0) + \frac{d-\|\xi_n\|^2-\mathrm{tr}(IJ^{-1})}{2n} + o_p(1/n), \tag{4.27}
$$

$$
C_n = L_n(w_0) + \frac{d-\|\xi_n\|^2+\mathrm{tr}(IJ^{-1})}{2n} + o_p(1/n), \tag{4.28}
$$

$$
W_n = L_n(w_0) + \frac{d-\|\xi_n\|^2+\mathrm{tr}(IJ^{-1})}{2n} + o_p(1/n). \tag{4.29}
$$

Proof. This theorem is obtained by combining Theorem 3 and Lemma 4.19.

$\square$

Theorem 7. *(Expectation of regular asymptotic theory) The asymptotic expansions of the generalization loss, the training loss, the cross validation loss, and WAIC are given by*

$$\mathbb{E}[G_n] = L(w_0) + \frac{d}{2n} + o(1/n), \tag{4.30}$$

$$\mathbb{E}[T_n] = L(w_0) + \frac{d - 2\mathrm{tr}(IJ^{-1})}{2n} + o(1/n), \tag{4.31}$$

$$\mathbb{E}[C_n] = L(w_0) + \frac{d}{2n} + o(1/n), \tag{4.32}$$

$$\mathbb{E}[W_n] = L(w_0) + \frac{d}{2n} + o(1/n). \tag{4.33}$$

Proof. Since $\mathbb{E}[G_{n-1}] = \mathbb{E}[C_n]$, $\mathbb{E}[\|\xi_n\|^2] = \mathrm{tr}(IJ^{-1}) + o(1)$. Then by Lemma 12, we obtain Theorem 7. $\square$

Remark 24. The equation $\mathbb{E}[\|\xi_n\|^2] = \mathrm{tr}(IJ^{-1}) + o(1)$ can be proved by two methods. The first proof is derived from the definition

$$\xi_n = J^{-1/2} \frac{1}{\sqrt{n}} \sum_{i=1}^{n} \nabla f(X_i, w_0).$$

The second is given by

$$\mathbb{E}[G_{n-1}] = \mathbb{E}[C_n].$$

If $X_1, X_2, ..., X_n$ are independent, then both methods can be used. However, if they are not independent, then the second method cannot be used. In such a case, the first proof can be applied.

Remark 25. (1) By the theorem, the convergence in probability holds,

$$n(G_n - L(w_0)) + n(C_n - L_n(w_0)) \to d,$$

where d is the dimension of the parameter. That is to say, for a given triple $(q(x), p(x|w), \varphi(w))$, if $(C_n - L_n(w_0))$ is smaller then $(G_n - L_n(w_0))$ is larger. (2) In regular asymptotic theory, the functional variance V_n is given by

$$V_n = \frac{1}{n} \sum_{i=1}^{n} \mathbb{V}_w[\log p(X_i|w)]$$

$$= \frac{\mathrm{tr}(IJ^{-1})}{n} + o_p(1/n).$$

4.4   Proof of Asymptotic Expansions

In this section, we prove Lemma 12, which is the base of the asymptotic expansion. The random variable we study is

$$Y = \sup_{|\alpha| \leq 1} \mathbb{E}_X \left[\frac{n^{k/2} \, \mathbb{E}_w[|g(X, w)|^k \exp(-\alpha f(X, w))]}{\mathbb{E}_w[\exp(-\alpha f(X, w))]} \right].$$

Then it is sufficient to prove that $\mathbb{E}[|Y|^{1+\varepsilon}] < \infty$ for an arbitrary $\varepsilon > 0$. Let $W_1 = \{w; \|w - w_0\| < n^{-2/5}\}$ and $W_2 = W \setminus W_1$, and

$$
\begin{align}
Y_1 &= \mathbb{E}_w[\exp(-\alpha f(X, w))]_{\{W_1\}}, & (4.34) \\
Y_2 &= \mathbb{E}_w[\exp(-\alpha f(X, w))]_{\{W_2\}}, & (4.35) \\
Y_3 &= n^{k/2} \mathbb{E}_w[|g(X, w)|^k \exp(-\alpha f(X, w))]_{\{W_1\}}, & (4.36) \\
Y_4 &= n^{k/2} \mathbb{E}_w[|g(X, w)|^k \exp(-\alpha f(X, w))]_{\{W_2\}}, & (4.37)
\end{align}
$$

where $\mathbb{E}[f(X)]_{\{S\}}$ is the expected value with restriction a set S. Then Y can be rewritten as

$$Y = \sup_{|\alpha| \leq 1} \mathbb{E}_X \left[\frac{Y_3 + Y_4}{Y_1 + Y_2} \right].$$

Since $Y_1, Y_2, Y_3, Y_4 > 0$,

$$0 < \frac{Y_3 + Y_4}{Y_1 + Y_2} \leq \frac{Y_3}{Y_1} + \min\left\{ \frac{Y_4}{Y_1}, \frac{Y_4}{Y_2} \right\}.$$

Firstly, let us study Y_3/Y_1.

Lemma 16. *Let $a, c, D, n > 0$ and $d \geq 0$ be arbitrary real constants, and $f_1(u)$ and $f_2(u)$ be real-valued continuous functions of $u \in [0, D] \subset \mathbb{R}$ which are differentiable in $(0, D)$. For nonnegative integer $m \geq 0$, Z_m is defined by*

$$Z_m = \int_0^D (au^{2c})^{m/2} u^d \exp(-nau^{2c} + \sqrt{n}au^c f_1(u) + f_2(u))du.$$

Then

$$\frac{Z_m}{Z_0} \leq \frac{2^{m/2} A^m + 2^m B^{m/2}}{n^{m/2}},$$

where, by using a definition $q = (m - 2)c + d + 1$,

$$
\begin{align}
A &= (c \sup_u |f_1(u)| + D \sup_u |f_1'(u)|)/(2c), & (4.38) \\
B &= (q + D \sup_u |f_2'(u)|)/(2c). & (4.39)
\end{align}
$$

Proof. For the proof we define

$$H(u) = \exp(\sqrt{n}au^c f_1(u) + f_2(u)). \tag{4.40}$$

Note that $(2c - 1) + q = cm + d$. By using partial integral,

$$
\begin{aligned}
Z_m &= \int_0^D \{au^{2c-1} \exp(-nau^{2c})\}\{u^q H(u)\} du \\
&= -\frac{1}{2cn} \int_0^D \partial_u\{\exp(-nau^{2c})\}\{u^q H(u)\} du \\
&= -\frac{1}{2cn} \Big[\exp(-nau^{2c})\{u^q H(u)\}\Big]_0^D \\
&\quad +\frac{1}{2cn} \int_0^D \{\exp(-nau^{2c})\}\partial_u\{u^q H(u)\} du \\
&\leq \frac{1}{2cn} \int_0^D \exp(-nau^{2c})\partial_u\{u^q H(u)\} du. \tag{4.41}
\end{aligned}
$$

Then by the assumption,

$$
\begin{aligned}
\partial_u\{u^q H(u)\} &= u^{q-1} H(u)\{q + u f_2'(u) \\
&\quad +c\sqrt{n}au^c f_1(u) + \sqrt{n}au^{c+1} f_1'(u)\} \\
&\leq u^{q-1} H(u)\{q + D \sup |f_2'| \\
&\quad +\sqrt{n}au^c(c \sup |f_1| + D \sup |f_1'|)\}. \tag{4.42}
\end{aligned}
$$

By applying eq.(4.42) to eq.(4.41) and using definition of Z_{m-2} and Z_{m-1},

$$Z_m \leq \frac{Z_{m-2}}{2cn}(q + D \sup |f_2'|) + \frac{Z_{m-1}}{2c\sqrt{n}}(c \sup |f_1| + D \sup |f_1'|).$$

By using definitions of eq.(4.38) and eq.(4.39), we obtain an inequality,

$$\frac{Z_m}{Z_0} \leq \frac{A}{\sqrt{n}}\frac{Z_{m-1}}{Z_0} + \frac{B}{n}\frac{Z_{m-2}}{Z_0}.$$

By Cauchy-Schwarz inequality,

$$\frac{Z_{m-1}}{Z_0} \leq \left(\frac{Z_m}{Z_0}\frac{Z_{m-2}}{Z_0}\right)^{1/2}.$$

Hence

$$
\begin{aligned}
\frac{Z_m}{Z_0} &\leq \frac{A}{\sqrt{n}}\left(\frac{Z_m}{Z_0}\frac{Z_{m-2}}{Z_0}\right)^{1/2} + \frac{B}{n}\frac{Z_{m-2}}{Z_0} \\
&= \left(\frac{Z_m}{Z_0} \cdot \frac{A^2}{n}\frac{Z_{m-2}}{Z_0}\right)^{1/2} + \frac{B}{n}\frac{Z_{m-2}}{Z_0} \\
&\leq \frac{1}{2}\left(\frac{Z_m}{Z_0} + \frac{A^2}{n}\frac{Z_{m-2}}{Z_0}\right) + \frac{B}{n}\frac{Z_{m-2}}{Z_0}.
\end{aligned}
$$

It follows that

$$\frac{Z_m}{Z_0} \le \frac{A^2 + 2B}{n} \frac{Z_{m-2}}{Z_0}.$$

Hence

$$\frac{Z_m}{Z_0} \le \frac{(A^2 + 2B)^{m/2}}{n^{m/2}} \le \frac{2^{m/2} A^m + 2^m B^{m/2}}{n^{m/2}},$$

which completes the lemma.

$\square$

Lemma 17. *Let Y_1 and Y_3 be random variables defined by eq.(4.34) and eq.(4.36) respectively. We define*

$$T(X) = \sup_{w \in W_1} \|J^{-1/2} \nabla g(X, w)\|.$$

Then

$$\frac{Y_3}{Y_1} \le T(X)^k (2^k A^{2k} + 2^{2k} B^k),$$

where

$$A \le c_1 \|\nabla \eta_n(w_0)\| \qquad (4.43)$$
$$B \le q + c_2 n^{-2/5} (\sup_w |\nabla_w f(X, w)| + \sup_w |\nabla_w \delta_n(w)|). \qquad (4.44)$$

Here by using $q = (2k - 2) + d$, and $c_1, c_2 > 0$ are constants determined by J. Hence $\mathbb{E}[(Y_3/Y_1)^{1+\varepsilon}] < \infty$.

Proof. By the definitions,

$$\frac{Y_3}{Y_1} = n^{k/2} \frac{\int_{W_1} |f(X, w)|^k \exp(-n K_n(w) - \alpha f(X, w)) \varphi(w) dw}{\int_{W_1} \exp(-n K_n(w) - \alpha f(X, w)) \varphi(w) dw}$$

where

$$n K_n(w) = \frac{n}{2} \|J^{1/2}(w - w_0)\|^2 - \nabla \eta_n(w_0) \cdot (w - w_0) + \delta_n(w).$$

By $J^{1/2}(w - w_0) = r\theta$ where θ is the generalized polar coordinate with $|\theta| = 1$. Then

$$\|J^{1/2}(w - w_0)\|^2 = r^2.$$

By

$$T(X) = \sup_{w \in W_1} \|J^{-1/2}\nabla f(X, w)\|$$

it follows that

$$|f(X, w)| \le rT(X).$$

Hence

$$\frac{Y_3}{Y_1} \le T(X)^k \frac{\int_{W_1} (nr^2)^{k/2} \exp(-nr^2 + \sqrt{n}r f_1(\theta) + f_2(r, \theta)) r^{d-1} dr d\theta}{\int_{W_1} \exp(-nr^2 + \sqrt{n}r f_1(\theta) + f_2(r, \theta)) r^{d-1} dr d\theta},$$

where

$$\begin{aligned} f_1(\theta) &= J^{-1/2}\nabla\eta_n(w_0) \cdot \theta \\ f_2(r, \theta) &= -\alpha f(X, r, \theta) + \log\varphi(r, \theta) - \delta_n(r, \theta) \end{aligned}$$

By applying Lemma 16,

$$\frac{Y_3}{Y_1} \le T(X)^k (2^k A^{2k} + 2^{2k} B^k),$$

where, by using $q = (2k - 2) + d$, $D = \|J^{1/2}\|n^{-2/5}$

$$\begin{aligned} A &= \sup_{\theta} |f_1(\theta)|/2 \\ B &= (q + D \sup_{(\ell, \theta)} |\partial_r f_2(r, \theta)|)/2, \end{aligned}$$

where we used the inequality

$$\left|\frac{\partial f}{\partial r}\right| = \left|\frac{\partial f}{\partial w} \cdot \frac{\partial w}{\partial r}\right| \le |\nabla f| \|J^{-1/2}\|.$$

$\square$

Lemma 18. *Let*

$$\overline{f}(X) = \sup_{w \in W} |f(X, w)|, \quad \overline{g}(X) = \sup_{w \in W} |g(X, w)|.$$

Then

$$\begin{aligned} \frac{Y_4}{Y_1} &\le \overline{g}(X)^k \exp(2\alpha\overline{f}(X)) \exp(-(J_1/4)n^{1/5} + (k/2)\log n + \gamma_n), \\ \frac{Y_4}{Y_2} &\le n^{k/2}\overline{g}(X)^k. \end{aligned}$$

Proof. Then by definition,

$$Y_1 \geq \exp(-\alpha\overline{f}(X))Z_n^{(1)}.$$
$$Y_4 \leq n^{k/2}\overline{g}(X)^k \exp(\alpha\overline{f}(X))Z_n^{(2)}.$$

Hence

$$Y_4/Y_1 \leq n^{k/2}\overline{g}(X)^k \exp(2\alpha\overline{f}(X))Z_n^{(2)}/Z_n^{(1)},$$
$$\leq n^{k/2}\overline{g}(X)^k \exp(2\alpha\overline{f}(X)) \exp(-(J_1/4)n^{1/5} + \gamma_n).$$

On the other hand,

$$Y_4/Y_2 \leq n^{k/2}\overline{g}(X)^k$$

is derived from the definition of $\overline{f}(X)$. $\square$

Lemma 19. *Let A_n and B_n be random variables. Assume that*

$$M = \sup_n \mathbb{E}[(|A_n| + |B_n|)^{m+\ell}] < \infty$$

and that $\{a_n > 0\}$ is an increasing sequence of real values. Then

$$\mathbb{E}[|A_n|^m \min\{\exp(A_n + B_n - a_n), n^{m/2}\}]$$
$$\leq \mathbb{E}[|A_n|^m] + Mn^{m/2}/(a_n)^\ell. \tag{4.45}$$

Proof. By the assumption

$$M = \mathbb{E}[(|A_n| + |B_n|)^{m+\ell}]$$

is finite. Hence

$$M \geq (a_n)^\ell \mathbb{E}[(|A_n|| + |B_n|)^m]_{\{|A_n|+|B_n|\geq a_n\}}.$$

Let E_n be the left hand side of eq.(4.45).

$$E_n \leq \mathbb{E}[|A_n|^m]_{\{|A_n|+|B_n|<a_n\}} + \mathbb{E}[n^{k/2}|A_n|^m]_{\{|A_n|+|B_n|\geq a_n\}}$$
$$\leq \mathbb{E}[|A_n|^m] + Mn^{m/2}/(a_n)^\ell,$$

which completes the lemma. $\square$

Let us prove Lemma 12

Proof. (Proof of Lemma 12). By Lemma 17, $\mathbb{E}[|Y_3/Y_1|^{1+\varepsilon}] < \infty$. In Lemma 19, by putting $m = k + \varepsilon$, $A_n = \sup_w |f(X, w)|$, $B_n = \gamma_n$, and $a_n = (J_1/4)n^{1/5} - (k/2)\log n$, and $\ell = 5m/2$,

$$\mathbb{E}\left[\min\left\{\left(\frac{Y_4}{Y_1}, \frac{Y_4}{Y_2}\right)\right\}^{1+\varepsilon}\right] < \infty,$$

which completes the lemma. $\square$

4.5 Point Estimators

In this section, we study other statistical estimation methods when a true distribution is regular for a statistical model.

Definition 12. The maximum likelihood estimator w_{ML}, the maximum *a posteriori* estimator w_{MAP}, and the posterior mean estimator w_{PM} are defined by

$$w_{ML} = \arg\max_{w \in W} \prod_{i=1}^{n} p(X_i|w), \tag{4.46}$$

$$w_{MAP} = \arg\max_{w \in W} \varphi(w) \prod_{i=1}^{n} p(X_i|w), \tag{4.47}$$

$$w_{PM} = \mathbb{E}_w[w], \tag{4.48}$$

where "$\arg\max_{w \in W} f(w)$" is the parameter that maximizes $f(w)$. The procedures in which a true distribution $q(x)$ is estimated by $p(x|w_{ML})$, $p(x|w_{MAP})$, and $p(x|w_{PM})$ are respectively called the maximum likelihood, maximum *a posteriori*, and the posterior mean methods.

If a true distribution is regular for a statistical model, by using

$$nK_n(w) = \frac{n}{2}\|J^{1/2}(w - w_0)\|^2 - \nabla_n(w_0) \cdot (w - w_0) + o_p(1),$$

these three estimators are asymptotically equivalent,

$$w_{ML} = w_0 + \frac{J^{-1/2}\xi_n}{\sqrt{n}} + o_p(\frac{1}{\sqrt{n}}),$$

$$w_{MAP} = w_0 + \frac{J^{-1/2}\xi_n}{\sqrt{n}} + o_p(\frac{1}{\sqrt{n}}),$$

$$w_{PM} = w_0 + \frac{J^{-1/2}\xi_n}{\sqrt{n}} + o_p(\frac{1}{\sqrt{n}}).$$

Remark 26. (1) Even if a same prior distribution is employed, $p(x|w_{MAP})$ and $p(x|w_{PM})$ are different from the Bayesian estimation $\mathbb{E}_w[p(x|w)]$. In some books and papers, $p(x|w_{MAP})$ and $p(x|w_{PM})$ may be called 'Bayesian estimation' and $\mathbb{E}_w[p(x|w)]$ is called 'fully Bayesian estimation'.
(2) If a true distribution is not regular for a statistical model, then w_{ML}, w_{MAP}, and w_{PM} are not equivalent even asymptotically.

Lemma 20. *The generalization and training losses of the maximum likeli-hood method are defined by*

$$G_n(ML) = L(w_{ML}), \tag{4.49}$$
$$T_n(ML) = L_n(w_{ML}). \tag{4.50}$$

If a true model is regular for a statistical model,

$$G_n(ML) = L(w_0) + \frac{\|\xi_n\|^2}{2n} + o_p(\frac{1}{n}), \tag{4.51}$$

$$T_n(ML) = L_n(w_0) - \frac{\|\xi_n\|^2}{2n} + o_p(\frac{1}{n}). \tag{4.52}$$

The generalization and training losses of the maximum a posteriori method and posterior mean method are asymptotically equivalent to those of the maximum likelihood method.

Proof. The generalization and training losses by the maximum likelihood method are given by

$$G_n(ML) = L(w_0) + K(w_{ML}), \tag{4.53}$$
$$T_n(ML) = L_n(w_0) + K_n(w_{ML}). \tag{4.54}$$

By $J = \nabla^2 K(w_0)$,

$$
\begin{aligned}
K(w_{ML}) &= \frac{1}{2}(w_0 - w_{ML}) \cdot \nabla^2 K(w_0)(w_0 - w_{ML}) + o_p(\frac{1}{n}) \\
&= \frac{\|\xi_n\|^2}{2n} + o_p(\frac{1}{n})
\end{aligned}
$$

and by $\nabla K_n(w_0) = -(1/\sqrt{n})\nabla \eta_n(w_0) = -(1/\sqrt{n})J^{1/2}\xi_n$, and

$$\nabla^2 K_n(w_0) = \nabla^2 K(w_0) + o_p(1),$$

it follows that

$$
\begin{aligned}
K_n(w_{ML}) &= (w_0 - w_{ML}) \cdot \nabla K_n(w_0) \\
&\quad + \frac{1}{2}(w_0 - w_{ML}) \cdot \nabla^2 K_n(w_0)(w_0 - w_{ML}) + o_p(\frac{1}{n}) \\
&= -\frac{\|\xi_n\|^2}{2n} + o_p(\frac{1}{n}).
\end{aligned}
$$

$\square$

The generalization and training losses of the maximum it a posteriori and the posterior mean methods have the same asymptotic behaviors as the maximum likelihood method. If a true distribution is realizable by a statistical model, then $I = J$, resulting that these asymptotic losses are equivalent to those of Bayesian estimation in Theorem 6.

Remark 27. (Deviance and functional variance) In Bayesian estimation, the deviance Dev and functional variance V are defined by

$$\text{Dev} = -\frac{2}{n}\Big\{\sum_{i=1}^{n}\mathbb{E}_w[\log p(X_i|w)] - \sum_{i=1}^{n}\log p(X_i|\mathbb{E}_w[w])\Big\},$$

$$V = \frac{1}{n}\sum_{i=1}^{n}\Big\{\mathbb{E}[(\log p(X_i|w))^2] - \mathbb{E}[\log p(X_i|w)]^2\Big\}.$$

If a true distribution is regular for a statistical model and if n is sufficiently large,

$$\text{Dev} = 2\{\mathbb{E}_w[K_n(w)] - K_n(\mathbb{E}_w[w])\}$$

$$= 2\{\frac{d - \|\xi_n\|^2}{2n} - \frac{-\|\xi_n\|^2}{2n}\} + o_p(\frac{1}{n})$$

$$= \frac{d}{n} + o_p(\frac{1}{n}), \tag{4.55}$$

and

$$V = \mathcal{T}_n''(0) = \frac{\text{tr}(IJ^{-1})}{n} + o_p(1/n). \tag{4.56}$$

Note that eq.(4.55) and eq.(4.56) hold even if a true distribution is not realizable by a statistical model.

In practical applications, we do not know the true distribution, hence the optimal parameter w_0 is unknown. Let $\hat{w}$ be the MAP estimator. Let us introduce a numerical calculation method for the free energy,

$$F_n = -\log \int \exp(-n\mathcal{L}(w))dw.$$

Here $\mathcal{L}(w)$ is the sum of the log likelihood function and log prior,

$$\mathcal{L}(w) = nL_n(w) - \log \varphi(w).$$

By the regularity condition,

$$n\mathcal{L}(w) \approx nL_n(\hat{w}) + \frac{n}{2}(w - \hat{w})J_n(w - \hat{w}) + o_p(1),$$

where we used a notation, $J_n = \nabla^2 L_n(\hat{w})$. It follows that

$$F_n = nL_n(\hat{w}) + \frac{d}{2}\log n - \frac{d}{2}\log(2\pi) + \frac{1}{2}\log \det J_n - \log \varphi(\hat{w}) + o_p(1). \quad (4.57)$$

The information criterion BIC is defined by

$$\mathrm{BIC} = \mathrm{nL}_n(\hat{w}) + \frac{d}{2}\log \mathrm{n}. \quad (4.58)$$

The cross validation and training losses are

$$C_n \approx L_n(\hat{w}) + \frac{d + \mathrm{tr}(I_n J_n^{-1})}{2n} + o_p(1/n), \quad (4.59)$$

$$T_n \approx L_n(\hat{w}) + \frac{d - \mathrm{tr}(I_n J_n^{-1})}{2n} + o_p(1/n), \quad (4.60)$$

where

$$I_n = \frac{1}{n}\sum_{i=1}^{n} \nabla f(X_i, \hat{w})(\nabla f(X_i, \hat{w}))^T.$$

4.6   Problems

1. Let $p(x|m, s)$ and $\varphi(m, s|\phi)$ be a statistical model and a prior given by eq.(2.1) and eq.(2.2), respectively. Show that the MAP estimator $(\hat{m}, \hat{s})$ is

$$\hat{m} = \hat{\phi}_2/\hat{\phi}_3,$$
$$\hat{s} = \hat{\phi}_3^2/(\hat{\phi}_1\hat{\phi}_3 - \hat{\phi}_2^2),$$

where $\hat{\phi}_1$, $\hat{\phi}_2$, and $\hat{\phi}_3$ are given by eq.(2.6), (2.7), and (2.8). The log loss function is given by

$$L_n(w) = \frac{1}{2}\log(2\pi) - \frac{\log s}{2} + \frac{s}{2n}\sum_{i=1}^{n}(X_i - m)^2.$$

Show that the matrix $I_n(m, s)$ is given by

$$(I_n)_{11}(m, s) = \frac{s^2}{n}\sum_{i=1}^{n}(X_i - m)^2,$$

$$(I_n)_{12}(m, s) = \frac{1}{2n}\sum_{i=1}^{n}(1 - s(X_i - m)^2)(X_i - m),$$

$$(I_n)_{22}(m, s) = \frac{1}{4n}\sum_{i=1}^{n}(1/s - (X_i - m)^2)^2,$$

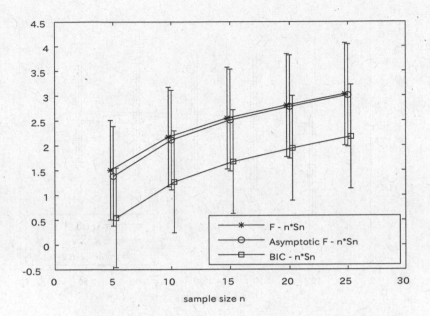

Figure 4.1: Free energy and its asymptotic form in normal distribution. Values $F_n - nS_n$ are compared with the asymptotic form and BIC. In a normal distribution, $F_n - nS_n$ can be approximated by its asymptotic form when $n \geq 10$. The difference between F_n and BIC is a constant order term.

and $(I_n)_{21}(m, s) = (I_n)_{12}(m, s)$. Also the matrix $J_n(m, s)$ is given by

$$(J_n)_{11}(m, s) = s,$$

$$(J_n)_{12}(m, s) = m - \frac{1}{n} \sum_{i=1}^{n} X_i,$$

$$(J_n)_{22}(m, s) = 1/(2s^2),$$

and $(J_n)_{21}(m, s) = (J_n)_{12}(m, s)$. A true parameter w_0 and a hyperparamter ϕ are determined as Example 5. The free energy or the minus log marginal likelihood F_n and the empirical entropy are given by eq.(2.9) and $L_n(w_0)$, respectively. The asymptotic form of F_n and BIC are given by eq.(4.57) and eq.(4.58), respectively. In Figure 4.1, $F_n - nS_n$ is compared with its asymptotic form and BIC, by using the numerical calculation. In a normal distribution, the free energy can be approximated by its asymptotic form when $n \geq 10$. The difference between F_n and BIC is a constant order term. In Figure 4.2, $\mathrm{tr}(I_n(\hat{w})J_n^{-1}(\hat{w}))$ is compared with nV and $n(C_n - T_n)$, where

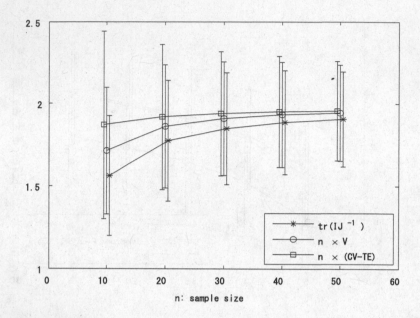

Figure 4.2: Comparison of $\text{tr}(IJ^{-1})$ with V in normal distribution. Values $\text{tr}(I_n(\hat{w})J_n^{-1}(\hat{w}))$ are compared with nV, and $n(CV - TE)$, where V is the functional variance, $CV - TE$ is the difference between the cross validation error and the training error in a simple normal distribution.

V and $C_n - T_n$ are the functional variance and the difference between the cross validation and training loss. In this case, the variance of the $n(C_n - T_n)$ is larger than the others. The values $n(C_n - T_n)$ and nV are approximated by $\text{tr}(I_n J_n^{-1})$ when $n \geq 40$. From the numerical point of view, the precise approximation of the free energy does not ensure that of the generalization loss.

2. A statistical model is defined by

$$p(x|a, b) = (1 - a)N(x) + aN(x - b),$$

where $N(x)$ is the probability density of the standard normal distribution. A true density is set as $q(x) = p(x|0.5, 1)$ and the uniform prior for (a, b) is given by $[0, 1] \times [0, 2]$. Then the true distribution is realizable by and regular for a statistical model and the maximum *a posteriori* estimator is equal to the maximum likelihood estimator. The free energy or the minus

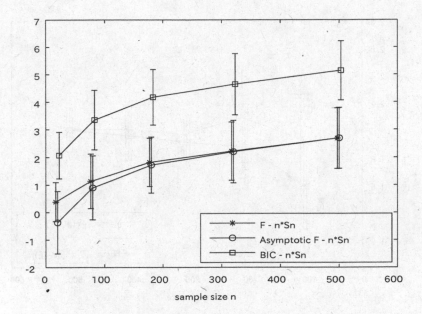

Figure 4.3: Free energy and its asymptotic form in a normal mixture. Values $F_n - nS_n$ are compared with its asymptotic form and BIC. In a normal mixture, $F_n - nS_n$ can be approximated by its asymptotic form when $n \geq$ 200. The difference between F_n and BIC is a constant order term.

log marginal likelihood F_n and the empirical entropy are given by eq.(2.9) and $L_n(w_0)$, respectively. The asymptotic form of F_n and BIC are given by eq.(4.57) and eq.(4.58), respectively. In this experiment, the integration of a function $f(a,b)$ over the parameter (a,b) is performed by the Riemann sum

$$\int_0^1 da \int_0^2 f(a,b)dadb = \frac{1}{2N^2} \sum_{j=1}^N \sum_{k=1}^N f(j/N - 1/2, 2k/N - 1).$$

In Figure 4.3, $F_n - nS_n$ is compared with its asymptotic form and BIC, by using the numerical calculation. In a normal mixture, the free energy can be approximated by its asymptotic form when $n \geq 200$. The difference between F_n and BIC is a constant order term. In Figure 4.4, $ntr(I_n(\hat{w})J_n^{-1}(\hat{w}))$ is numerically compared with nV where V is the functional variance. The horizontal and vertical axes show the sample size and the average and standard deviation for 100 sample sets X^n, respectively. The true parameter is the regular point for the statistical model. Let $M(x) = N(x) - N(x-b)$ and

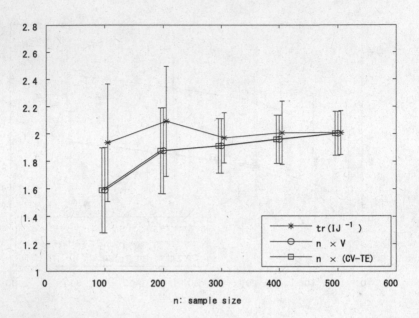

Figure 4.4: Comparison with $\operatorname{tr}(IJ^{-1})$ with V in normal mixture. Values for $\operatorname{tr}(I_n(\hat{w})J_n^{-1}(\hat{w}))$ are compared with nV, and $n(CV - GE)$, where V is the functional variance, CV is the cross validation error, and TE is the training error in a normal mixture.

$p(x) = p(x|a, b)$. Show that

$$(I_n)_{11}(a, b) = \frac{1}{n}\sum_{i=1}^{n} M(X_i)^2/p(X_i)^2,$$

$$(I_n)_{12}(a, b) = \frac{1}{n}\sum_{i=1}^{n} aM(X_i)N'(x - b)/p(X_i)^2,$$

$$(I_n)_{22}(a, b) = \frac{1}{n}\sum_{i=1}^{n} \{aN'(x - b)\}^2/p(X_i)^2,$$

and $(I_n)_{21}(a, b) = (I_n)_{12}(a, b)$. Also show that the matrix $J_n(a, b)$ is given

by

$$(J_n)_{11}(a,b) \;=\; \frac{1}{n}\sum_{i=1}^{n} M(X_i)^2/p(X_i)^2,$$

$$(J_n)_{12}(a,b) \;=\; \frac{1}{n}\sum_{i=1}^{n} N'(x-b)\{p(X_i)+aM(X_i)\}/p(X_i)^2,$$

$$(J_n)_{22}(a,b) \;=\; \frac{1}{n}\sum_{i=1}^{n}\{-aN''(X_i-b)/p+(aN'(X_i-b))^2/p^2\},$$

and $(J_n)_{21}(a,b) = (J_n)_{12}(a,b)$. The varinace of $\mathrm{tr}(I_n(\hat{w})J_n^{-1}(\hat{w}))$ is larger than those of nV and $n(C_n - T_n)$. The dimension of the parameter space is two, and the true parameter is a regular point for the log loss function. However, the regular theory needs $n \geq 500$. From the numerical point of view, the precise approximation of the free energy does not ensure that of the generalization loss.

3. A neural network is defined by a conditional density

$$p(y|x,a,b) = \frac{1}{\sqrt{2\pi}}\exp(-\frac{1}{2}(y-a\tanh(bx))^2).$$

A probability density of x is set as the uniform distribution of $[-2,2]$. A true conditional density is set as $p(y|x,1,1)$, and the uniform prior for (a,b) is given by $[0,2] \times [0,2]$. Then the true distribution is realizable by and regular for a statistical model and the maximum *a posteriori* estimator is equal to the maximum likelihood estimator. The value $n\mathrm{tr}(I_n(\hat{w})J_n^{-1}(\hat{w}))$ is numerically compared with nV where V is the functional variance. The experimental result is shown in Figure 4.5. The horizontal and vertical axes show the sample size and the average and standard deviation for 100 sample sets (X^n, Y^n), respectively. In this case the true parameter is the regular point for the statistical model. Let

$$\begin{aligned}
Z_0(x) &= \tanh(bx),\\
Z_1(x) &= (1-Z_0(x)^2)x,\\
Z_2(x) &= -2Z_0(x)Z_1(x)x.
\end{aligned}$$

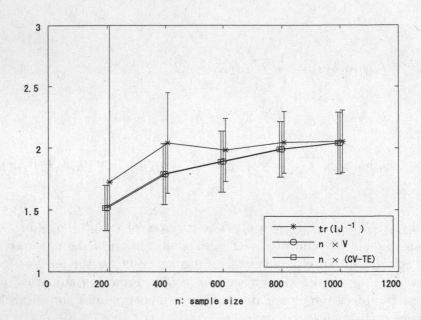

Figure 4.5: Comparison with $\mathrm{tr}(IJ^{-1})$ with V in neural network. Values for $\mathrm{tr}(I_n(\hat{w})J_n^{-1}(\hat{w}))$ are compared with nV, and $n(CV - GE)$, where V is the functional variance, CV is the cross validation error, and TE is the training error in a neural network.

$$(I_n)_{11}(a,b) \;=\; \frac{1}{n}\sum_{i=1}^{n}(aZ_0(X_i) - Y_i)^2 Z_0(X_i)^2,$$

$$(I_n)_{12}(a,b) \;=\; \frac{1}{n}\sum_{i=1}^{n}a(aZ_0(X_i) - Y_i)^2 Z_0(X_i)Z_1(X_i),$$

$$(I_n)_{22}(a,b) \;=\; \frac{1}{n}\sum_{i=1}^{n}a^2(aZ_0(X_i) - Y_i)^2 Z_1(X_i)^2,$$

and $(I_n)_{21}(a,b) = (I_n)_{12}(a,b)$. Also show that the matrix $J_n(a,b)$ is given

by

$$(J_n)_{11}(a,b) = \frac{1}{n}\sum_{i=1}^{n} Z_0(X_i)^2,$$

$$(J_n)_{12}(a,b) = \frac{1}{n}\sum_{i=1}^{n} \{2aZ_0(X_i)Z_1(X_i) - Y_iZ_1(X_i)\},$$

$$(J_n)_{22}(a,b) = \frac{1}{n}\sum_{i=1}^{n} \{a^2Z_1(X_i)^2 + a(aZ_0(X_i) - Y_i)Z_2(X_i)\},$$

and $(J_n)_{21}(a,b) = (J_n)_{12}(a,b)$. The varinace of $\text{tr}(I_n(\hat{w})J_n^{-1}(\hat{w}))$ is largar than those of nV and $n(C_n - T_n)$. Tthe dimension of the parameter space is two, and the true parameter is a regular point for the log loss function, however, the regular theory needs $n \geq 1000$. In practical applications, neural networks which have many parameters are employed, hence the regular theory does not hold in all cases. However, the general theory introduced in the following chapters holds.

Chapter 5

Standard Posterior Distribution

If a true distribution is regular for a statistical model and if the posterior distribution can be approximated by a normal distribution, the difference between Bayesian and maximum likelihood estimations is not so large. However, the posterior distributions are often far from any normal distribution, showing that Bayesian estimation gives the more accurate inference than other estimation methods. In this chapter we study the case when the posterior density $p(w)$ is asymptotically given by

$$p(w) \propto \exp\left(-n\, w_1^{2k_1}\, w_2^{2k_2}\, \cdots\, w_d^{2k_d}\right).$$

It might seem that such a posterior density appears in a special case. However, in the next chapter, we show that most posterior densities are mathematically equivalent to this function. Therefore, the results of this chapter are the universal laws of Bayesian statistics.

This chapter consists of the following parts.

(1) A standard form is introduced and the real log canonical threshold is defined.

(2) Asymptotic property of a state density function is derived.

(3) Asymptotic behavior of the free energy or the minus log marginal likelihood is represented by the real log canonical threshold.

(4) By using the renormalized posterior distribution, mathematical laws among the generalization loss, cross validation loss, training loss, and WAIC are established.

(5) If random variables are conditionally independent, the relation between the generalization and cross validation losses does not hold. However, the

mathematical theorem which connects the generalization loss and WAIC can be proved.

5.1   Standard Form

In order to define a standard form, we need a representation of a multi-index.

Definition 13. (Multi-index) A d-dimensional multi-index k is defined by

$$k = (k_1, k_2, ..., k_d).$$

For a multi-index k, notations $k > 0$ and $k \geq 0$ are defined as follows.

$$k \geq 0 \Longleftrightarrow k_1 \geq 0, k_2 \geq 0, ..., k_d \geq 0.$$

and

$$k > 0 \Longleftrightarrow k \geq 0, \text{ there exists } j \text{ such that } k_j > 0.$$

Let $k \geq 0$. For a given variable $w = (w_1, w_2, ..., w_d) \in \mathbb{R}^d$ we define

$$w^k = w_1^{k_1} w_2^{k_2} \cdots w_d^{k_d},$$

where $0^0 = 1$.

Example 23. Let $d = 5$. For multi-indexes,

$$
\begin{aligned}
k &= (3, 6, 7, 0, 0), \\
h &= (1, 0, 0, 8, 0),
\end{aligned}
$$

and a variable $w = (w_1, w_2, w_3, w_4, w_5)$,

$$
\begin{aligned}
w^k &= w_1^3 w_2^6 w_3^7, \\
w^h &= w_1 w_4^8.
\end{aligned}
$$

Note that, if $h = (0, 0, 0, 0, 0)$, then

$$w^h = 1.$$

The following definition is the concept of the standard form. Let $q(x)$ and $p(x|w)$ be a true distribution and a statistical model, respectively.

Definition 14. (Standard form) Let $x \in \mathbb{R}^N$ and $w \in W \subset \mathbb{R}^d$. We study a statistical model $p(x|w)$ of x for a given parameter w. Assume that W is a compact set which is a closure of some open set. Also we assume that W contains the origin and that W_0 is the set of parameters which minimizes the average log loss function. The log density ratio function and its average are respectively denoted by

$$f(x, w) = \log \frac{p(x|w_0)}{p(x|w)}, \tag{5.1}$$

$$K(w) = \int q(x) f(x, w) dx. \tag{5.2}$$

We assume that the log density ratio function has relatively finite variance, hence $f(x, w)$ does not depend on the choice of $w_0 \in W_0$. A set of statistical model $p(x|w)$ and a prior $\varphi(w)$ is said to be a standard form if there exist functions $a(x, w)$ and $b(w)$ which satisfy

$$f(x, w) = w^k a(x, w), \tag{5.3}$$

$$K(w) = w^{2k}, \tag{5.4}$$

$$\varphi(w) = |w^h| \, b(w), \tag{5.5}$$

where both $k > 0$ and $h \geq 0$ are multi-indexes and $b(w) > 0$ in a neighborhood of the origin.

Remark 28. (Normal crossing function) If an average log density ratio function $K(w)$ is represented by w^{2k}, then it is called normal crossing. It might seem that a very special set of a statistical model and a prior has a standard form. However, in the next section, we show almost all statistical models and priors such as a normal mixture and a neural network can be made to be standard forms by using an algebraic geometrical transform of the parameter set. Therefore the statistical theory of this chapter holds for such statistical models and priors.

By the definition and

$$K(w) = \int q(x) f(x, w) dx,$$

it follows that

$$w^k = \int q(x) a(x, w) dx.$$

The set of optimal parameters $W_0 = \{w \in W; K(w) = 0\}$ is

$$W_0 = \{w \in W; w^{2k} = 0\},$$

which is equal to

$$W_0 = \bigcup_{j:k_j>0} \{w \in W \; ; \; w_j = 0\},$$

where $\bigcup_{j:k_j>0}$ shows the union of sets for all j such that $k_j > 0$.

Example 24. Let $x, w \in \mathbb{R}^1$ and $W = [-1, 1]$. A statistical model and a prior are

$$p(x|w) \;\; = \;\; \frac{1}{\sqrt{\pi}} \exp(-(x-w)^2),$$

$$\varphi(w) \;\; = \;\; |w|^2.$$

Assume that $q(x) = p(x|0)$. Then $w_0 = 0$ and

$$f(x, w) \;\; = \;\; w(w - x), \tag{5.6}$$

$$K(w) \;\; = \;\; w^2. \tag{5.7}$$

Therefore the set of a model and a prior is a standard form with $k = 1, h = 2$ and

$$a(x, w) \;\; = \;\; w - x, \tag{5.8}$$

$$b(w) \;\; = \;\; 1. \tag{5.9}$$

Example 25. Let $-1 \le x \le 1$, $y \in \mathbb{R}$, and $W = \{(s, t) \in \mathbb{R}^2; -1 \le s, t \le 1\}$. A statistical model and a prior are defined by

$$p(x, y|s, t) \;\; = \;\; \frac{1}{2\sqrt{2\pi}} \exp\left\{-\frac{1}{2}(y - sF(x, t))^2\right\},$$

$$\varphi(s, t) \;\; = \;\; 1/4,$$

where

$$F(x, t) = \sqrt{2}\frac{\sqrt{3}x + t}{\sqrt{t^2 + 1}}.$$

Therefore $p(x|s, t)$ is the uniform distribution on $[-1, 1]$. Assume that $q(x, y) = p(x, y|0, 0)$. Then

$$f(x, s, t) \;\; = \;\; s^2 F(x, t)^2/2 - syF(x, t), \tag{5.10}$$

$$K(s, t) \;\; = \;\; s^2. \tag{5.11}$$

Hence

$$a(x, s, t) \;\; = \;\; sF(x, t)^2/2 - yF(x, t), \tag{5.12}$$

$$b(s, t) \;\; = \;\; 1/4. \tag{5.13}$$

Therefore the set of a model and a prior is a standard form with $k = (1, 0)$, $h = (0, 0)$.

Example 26. Let $x = (y, z), w = (s, t) \in \mathbb{R}^2$ and

$$W = \{w = (s, t) \; ; \; s^2 + t^2 \le 1\}.$$

A statistical model and a prior are

$$p(y, z | s, t) = \frac{1}{\pi} \exp(-(y - s)^2 - (z - t)^2),$$
$$\varphi(s, t) = \frac{1}{\pi}.$$

Assume that $q(x) = p(x | 0, 0)$. Then $w_0 = (0, 0)$ and

$$f(x, w) = s^2 + t^2 - 2ys - 2zt,$$
$$K(w) = s^2 + t^2.$$

This is not a standard form. By using a polar coordinate (r, θ),

$$s = r \cos \theta,$$
$$t = r \sin \theta,$$

where $0 \le r \le 1$ and $0 \le \theta < 2\pi$, the model and prior are rewritten as

$$f(x, r, \theta) = r(r - 2y \cos \theta - 2z \sin \theta),$$
$$K(r, \theta) = r^2,$$
$$\varphi(w) dw = \frac{r}{\pi} dr d\theta.$$

Therefore the pair of a model and a prior is a standard form with $k = (1, 0), h = (1, 0)$ and

$$a(x, r, \theta) = r - 2y \cos \theta - 2z \sin \theta,$$
$$b(r, \theta) = \frac{1}{\pi}.$$

In this case, a true distribution $q(x)$ is regular for a statistical model $p(x | s, t)$, but not for $p(x | r, \theta)$.

Lemma 21. *Assume that a statistical model $p(x | w)$ has a relatively finite variance and that there exists a C^1-class function $K_0(w) > 0$ which satisfies*

$$f(x, w) = w^k a(x, w), \tag{5.14}$$
$$K(w) = w^{2k} K_0(w), \tag{5.15}$$
$$\varphi(w) = |w^h| b(w), \tag{5.16}$$

where $k > 0$ and $h \geq 0$ are multi-indexes and $b(w) > 0$ in a neighborhood of the origin. Since $k > 0$, we can assume $k_1 > 0$ without loss of generality. Also we assume that

$$W(dK_0) = \{w \in W \; ; \; K_0(w)^{1/2k_1} + \frac{w_1}{2k_1}K_0(w)^{1/2k_1-1}\frac{\partial K_0(w)}{\partial w_1} = 0\}$$

is a measure zero subset in W. Then by using

$$u_1 = K_0(w)^{1/(2k_1)}w_1,$$
$$u_2 = w_2,$$
$$\cdots$$
$$u_d = w_d,$$

the pair of a model and a prior is a standard form of u.

Proof. The absolute value of the determinant of the Jacobian matrix $|\partial u/\partial w|$ is equal to $|\partial u_1/\partial w_1|$ and

$$\frac{\partial u_1}{\partial w_1} = K_0(w)^{1/2k_1} + \frac{w_1}{2k_1}K_0(w)^{1/2k_1-1}\frac{\partial K_0(w)}{\partial w_1}.$$

Since $K_0(w) > 0$ is a C^1-class function and W is compact, there exist constants $A, B > 0$ such that

$$\min_{w \in W} K_0(w) > A,$$
$$\max_{w \in W}\left|\frac{\partial K_0(w)}{\partial w_1}\right| < B.$$

Therefore, in a neighborhood of the origin, $|\partial u_1/\partial w_1| > 0$. The map $w \mapsto u$ is one-to-one in the set $W \setminus W(\partial u_1/\partial w_1)$, where $W(\partial u_1/\partial w_1)$ is the set of all zero points of $|\partial u_1/\partial w_1|$. Thus its inverse function is well-defined in $W \setminus W(\partial u_1/\partial w_1)$, which is denoted by $w = g(u)$. Then

$$f(x, g(u)) = u^k a(x, g(u))/K_0(g(u))^{1/2},$$
$$\int q(x)f(x, g(u))dx = u^{2k},$$
$$\varphi(g(u))|g'(u)| = |u^h|b(g(u))|g'(u)|,$$

which shows the pair of a model and a prior is a standard form, where $|g'(u)|$ is the determinant of the Jacobian matrix of $w = g(u)$. $\square$

Example 27. Let $0 \le x \le 1$, $y \in \mathbb{R}$, and $W = \{(s,t) \in \mathbb{R}^2; 0 \le s,t \le 1\}$. A statistical model and a prior are defined by

$$p(x,y|s,t) = \frac{1}{\sqrt{2\pi}} \exp(-\frac{1}{2}(y - s\tanh(tx))^2),$$

$$\varphi(s,t) = 1.$$

Therefore $p(x|s,t)$ is the uniform distribution on $[0,1]$. Assume that $q(x,y) = p(x,y|0,0)$. Then

$$K(s,t) = \frac{s^2 t^2}{2} K_0(t),$$

where

$$K_0(t) = \int_0^1 \left(\frac{\tanh(tx)}{t}\right)^2 dx.$$

By the above lemma, the pair of a model and a prior is made to be a standard form by

$$s' = s,$$

$$t' = t(K_0(t)/2)^{1/2}.$$

In this case, a true distribution cannot be regular for a statistical model by any transform of parameters.

Example 28. Let $N(x)$ be a probability density function of the normal distribution whose average and standard deviation are zero and one respectively,

$$N(x) = \frac{1}{\sqrt{2\pi}} \exp(-\frac{x^2}{2}).$$

Let $x \in \mathbb{R}$ and $W = \{(s,t); 0 \le s \le 1, |t| < 1\}$. A statistical model and a prior are defined by

$$p(x|s,t) = (1-s)N(x) + sN(x-t),$$

$$\varphi(s,t) = 1/2.$$

Assume that a true distribution is $N(x)$. Then the set of optimal parameters is

$$W_0 = \{(s,t) \in W \; ; \; s = 0, \text{ or } t = 0\}.$$

The log density ratio function is

$$f(x,s,t) = \log \frac{N(x)}{(1-s)N(x) + sN(x-t)}$$

$$= -\log[1 + s\{e^{tx - t^2/2} - 1\}]$$

$$= -st\,(x - t/2)\,T(tx - t^2/2)\,S(s(e^{tx - t^2/2} - 1)),$$

where two analytic functions are defined by

$$
\begin{aligned}
S(x) &= \log(1+x)/x, \\
T(x) &= (e^x - 1)/x,
\end{aligned}
$$

with $T(0) = S(0) = 1$. Hence by

$$
a(x, s, t) = -(x - t/2)\, T(tx - t^2/2)\, S(s(e^{tx - t^2/2} - 1)),
$$

the pair of a model and a prior is a standard form. Note that

$$
\begin{aligned}
K(s, t) &= \int N(x)\{f(x, s, t) + \exp(-f(x, s, t)) - 1\}dx \\
&= \int N(x) f(x, s, t)^2 U(f(x, s, t))dx \\
&= (st)^2 \int N(x) a(x, s, t)^2 U(f(x, s, t))dx,
\end{aligned}
$$

where

$$
U(x) = \frac{(x + e^{-x} - 1)}{x^2}
$$

is an analytic function by defining $U(0) = 1/2$. In this case, a true distribution $N(x)$ cannot be regular for a statistical model by any transform of parameters.

Example 29. Let $0 \le x_1, x_2 \le 1$, $y \in \mathbb{R}$, and $W = \{(s, t_1, t_2) \in \mathbb{R}^3; 0 \le s \le 1, t_1^2 + t_2^2 \le 1\}$. A statistical model and a prior are defined by

$$
\begin{aligned}
p(x_1, x_2, y|s, t) &= \frac{1}{\sqrt{2\pi}} \exp(-\frac{1}{2}(y - s \tanh(t_1 x_1 + t_2 x_2))^2), \\
\varphi(s, t_1, t_2) &= 1/\pi.
\end{aligned}
$$

Therefore $p(x_1, x_2|s, t_1, t_2)$ is the uniform distribution on $[0, 1]^2$. Assume that $q(x_1, x_2, y) = p(x_1, x_2, y|0, 0, 0)$. By using

$$
\begin{aligned}
s &= s, \\
t_1 &= r \cos\theta, \\
t_2 &= r \sin\theta,
\end{aligned}
$$

where $0 \le r \le 1$, $-\pi \le \theta < \pi$,

$$
\begin{aligned}
K(s, r, \theta) &= \frac{s^2 r^2}{2} K_0(r, \theta), \\
(1/\pi)ds dt_1 dt_2 &= (r/\pi)ds dr d\theta,
\end{aligned}
$$

where

$$K_0(r,\theta) = \int_0^1 \int_0^1 \left(\frac{\tanh\{r(x_1 \cos\theta + x_2 \sin\theta)\}}{r} \right)^2 dx_1 dx_2.$$

The pair of a model and a prior is made a standard form.

Example 30. A neural network which has H hidden units is defined by

$$p(x, y|\{s_h, t_h\}) = \frac{1}{\sqrt{2\pi}} \exp(-\frac{1}{2}(y - \sum_{h=1}^{H} s_h \tanh(t_h x))^2).$$

In this case, in order to find a transform which makes the model a standard form, we need the method explained in the next chapter.

Remark 29. Almost all statistical models and priors used in practical applications can be made to be standard forms by choosing an appropriate function,

$$w = g(u).$$

Then a statistical model and a prior are rewritten as

$$p(x|w) = p(x|g(u)),$$
$$\varphi(w)dw = \varphi(u)|g'(u)|du,$$

where $|g'(u)|$ is the absolute value of the determinant of the Jacobian matrix,

$$|g'(u)| = \left| \det\left(\frac{\partial w}{\partial u} \right) \right|.$$

From the view point of Bayesian statistics, $(p(x|w), \varphi(w))$ is equivalent to $(p(x|g(u)), \varphi(w)|g'(u)|)$. In other words, the free energy and generalization, cross validation, and training losses are invariant by $w = g(u)$. A method to find the appropriate function $w = g(u)$ is discussed in the next chapter.

In order to study statistical estimation, we need additional mathematical conditions.

Definition 15. (Mathematical condition) (1) The set of parameters W is a compact set in $\mathbb{R}^d$ and the closure of its largest open set is equal to W.
(2) A statistical model is a standard form and $a(x, w)$ defined in eq.(5.3) satisfies that for an arbitrary $s > 0$ and arbitrary multi-index $k \geq 0$,

$$\int \sup_{w \in W} |(\partial/\partial w)^k a(x, w)|^s q(x) dx < \infty.$$

To construct statistical theory for a standard form, we need stochastic process because W_0 is not a single element.

Definition 16. Assume that a statistical model is a standard form. A stochastic process $\xi_n(w)$ is defined by

$$\xi_n(w) = \frac{1}{\sqrt{n}} \sum_{i=1}^{n} \{w^k - a(X_i, w)\}. \tag{5.17}$$

By this definition and the assumption that a statistical model is a standard form, it follows that

$$\mathbb{E}[\xi_n(w)] = 0, \tag{5.18}$$

$$\mathbb{E}[\xi_n(w)\xi_n(u)] = \mathbb{E}_X[a(X, w)a(X, u)] - w^k u^k. \tag{5.19}$$

On the other hand, the Gaussian process $\xi(w)$ on W that satisfies

$$\mathbb{E}_\xi[\xi(w)] = 0, \tag{5.20}$$

$$\mathbb{E}_\xi[\xi(w)\xi(u)] = \mathbb{E}_\xi[a(X, w)a(X, u)] - w^k u^k, \tag{5.21}$$

is uniquely determined. If $\xi_n(w)$ satisfies conditions of Definition 15,

$$\lim_{n \to \infty} \mathbb{E}[F(\xi_n)] = \mathbb{E}_\xi[F(\xi)]$$

holds for an arbitrary continuous and bounded functional $F(\ )$, where F is a function from a set of functions

$$\{f(w) ; \sup_{w \in W} |f(w)| < \infty\}$$

to $\mathbb{R}$. Moreover, for an arbitrary $s > 0$,

$$\lim_{n \to \infty} \mathbb{E}[\sup_{w \in W} |\xi_n(w)|^s] = \mathbb{E}_\xi[\sup_{w \in W} |\xi(w)|^s].$$

The mathematical background of these properties is explained in Section 10.4.

Example 31. Let $\{X_i; i = 1, 2, ..., n\}$ be a set of independent random variables which are subject to the uniform distribution on $[-1, 1]$. For a parameter a $(0 \le a \le 2\pi)$, an empirical process $\xi_n(a)$ is defined by

$$\xi_n(a) = \frac{1}{\sqrt{n}} \sum_{i=1}^{n} \sin(ax_i).$$

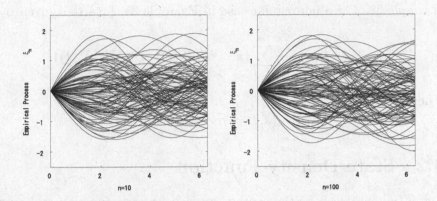

Figure 5.1: Examples of empirical processes are illustrated for $n = 10$ and $n = 100$. In the standard theory, the set of the optimal parameters W_0 consists of a union of several manifolds, empirical process theory is necessary.

For each a, $\xi_n(a)$ converges to a normal distribution whose average is zero and variance is $\mathbb{E}_X[\sin(aX)^2]$. See Figure 5.1. Moreover, $\xi_n(a)$ converges to a Gaussian process as a random process. Here a random process is a function-valued random variable.

Theorem 8. *Assume that a set of a statistical model and a prior is a standard form. The log likelihood ratio function is defined by*

$$K_n(w) = \frac{1}{n} \sum_{i=1}^{n} f(X_i, w).$$

Then

$$nK_n(w) = n\, w^{2k} - \sqrt{n}\, w^k\, \xi_n(w), \tag{5.22}$$

where $\xi_n(w)$ satisfies the convergence in distribution $\xi_n(w) \to \xi(w)$.

Proof. This theorem is shown by applying definitions $f(x, w) = w^k a(x, w)$, $K(w) = w^{2k}$, and eq.(5.17) to

$$nK_n(w) = nK(w) - n(K(w) - K_n(w)).$$

Then by using the empirical process theory in Section 10.4, we obtain the theorem. $\square$

Example 32. For a nonregular case in Example 25, $\xi_n(s,t)$ is given by

$$\xi_n(s,t) = \frac{1}{\sqrt{n}} \sum_{i=1}^{n} \{s - sF(X_i,t)^2/2 + Y_i F(X_i,t)\}.$$

Then

$$nK_n(s,t) = ns^2 - \sqrt{n}\, s\, \xi_n(s,t).$$

5.2   State Density Function

Let $K(w)$ and $\varphi(w)$ be the average log density ratio function and a prior of $w \in W \subset \mathbb{R}^d$ respectively. The posterior distribution has a form

$$\exp(-nK(w))\, \varphi(w),$$

which is the Laplace transform of the state density function

$$\delta(t - K(w))\, \varphi(w),$$

where $t > 0$. The asymptotic behavior for $n \to \infty$ corresponds to $t \to +0$, hence in this section we study the state density funciton and its asymptotic behavior.

Assume that a pair of a statistical model and a prior is a standard form. Then on the positive region of parameters,

$$w_1, w_2, ..., w_d \geq 0, \tag{5.23}$$

the state density function is

$$\delta(t - K(w))\varphi(w) = \delta(t - w^{2k})|w^h|b(w)\chi(w),$$

where $\chi(w)$ is the characteristic function of the positive region of parameters,

$$\chi(w) = \begin{cases} 1 & (w_1, w_2, ..., w_d \geq 0) \\ 0 & \text{(otherwise)} \end{cases}. \tag{5.24}$$

For general cases other than eq.(5.23), $\delta(t - K(w))\varphi(w)$ can be treated by using this case. See Remark 33.

Let us show that the asymptotic bahavior of the state density function is determined by the real log canonical threshold and its multiplicity determined by the multi-indexes k and h by the following definition.

Definition 17. (Real log canonical threshold and its multiplicity) Let $k > 0$ and $h \geq 0$ be d dimensional multi-indexes. Without loss of generality, we can assume that

$$\left\{ \frac{h_i + 1}{2k_i} \; ; \; i = 1, 2, ..., d \right\}$$

is a nondecreasing sequence of i, where, if $k_i = 0$, we define

$$\frac{h_i + 1}{2k_i} = +\infty.$$

Then by definition, there exist a positive real value $\lambda > 0$ and a positive integer m $(1 \leq m \leq d)$ such that

$$\lambda = \frac{h_1 + 1}{2k_1} = \frac{h_2 + 1}{2k_2} = \cdots = \frac{h_m + 1}{2k_m}$$

and

$$\lambda < \frac{h_j + 1}{2k_j} \quad (m + 1 \leq j \leq d).$$

Then the constants λ and m constitute a real log canonical threshold and its multiplicity. The redundant multi-index $\mu = (\mu_{m+1}, ..., \mu_d) \in \mathbb{R}^{d-m}$ is defined by

$$\mu_j = -2\lambda k_j + h_j \quad (j = m + 1, ..., d). \tag{5.25}$$

By the definition, $\mu_j > -1$. If $m = d$, then μ is the empty set.

Example 33. For the case

$$k = (3, 1, 4, 1, 2, 0, 0),$$
$$h = (2, 0, 3, 1, 5, 0, 1),$$

elements of the set $\{(h_i + 1)/(2k_i)\}$ are given by

$$\frac{2+1}{6}, \frac{0+1}{2}, \frac{3+1}{8}, \frac{1+1}{2}, \frac{5+1}{4}, \frac{0+1}{0}, \frac{1+1}{0},$$

which is a nondecreasing sequence. Therefore $\lambda = 1/2$ and $m = 3$, resulting that

$$(\mu_4, \mu_5, \mu_6, \mu_7) = (0, 3, 0, 1).$$

Definition 18. Let $k > 0$ and $h \geq 0$ be arbitrary multi-indexes. Let λ and m be the real log canonical threshold and its multiplicity, respectively. The constant $C(k, m)$ is defined by

$$C(k, m) = 2^m (m - 1)! \prod_{j=1}^{m} k_j. \tag{5.26}$$

We use a notation,

$$w = (w_\alpha, w_\beta),$$

where

$$\begin{aligned} w_\alpha &= (w_1, w_2, ..., w_m), \\ w_\beta &= (w_{m+1}, w_{m+2}, ..., w_d). \end{aligned}$$

Then a function $f(w)$ is rewritten as $f(w) = f(w_\alpha, w_\beta)$. A function (hyperfunction or distribution) $D(w)$ is defined by

$$D(w) = \frac{1}{C(k,m)} \, \delta(w_\alpha) \, |w_\beta^\mu| \, b(w)\chi(w), \qquad (5.27)$$

where μ is the redundant multi-index defined by eq.(5.25) and $\delta(w_\alpha)$ is defined by

$$\delta(w_\alpha) = \prod_{j=1}^{m} \delta(w_j).$$

If a function $\psi(w_j)$ is not continuous at $w_j = 0$, then

$$\int \delta(w_j)\psi(w_j)dw_j$$

is not defined. In such a case, we adopt a generalized delta function

$$\int \delta(w_j)\psi(w_j)dw_j = \lim_{\epsilon \to +0} \psi(\epsilon),$$

in the definition of $D(w)$.

Example 34. For k and m in Example 33, the constant $C(k, m)$ is

$$C(k,m) = 2^3(3-1)! \, 3 \cdot 1 \cdot 4 = 192.$$

By $m = 3$,

$$\begin{aligned} w_\alpha &= (w_1, w_2, w_3), \\ w_\beta &= (w_4, w_5, w_6, w_7). \end{aligned}$$

Then a function $f(w)$ is rewritten as $f(w) = f(w_\alpha, w_\beta)$. A function (hyperfunction or distribution) $D(w)$ is given by

$$D(w) = \frac{1}{192} \, \delta(w_1)\delta(w_2)\delta(w_3)|(w_4)^0(w_5)^3(w_6)^0(w_7)^1|b(w)\chi(w).$$

Since $\mu_j > -1$, $\int_W D(w)dw$ is a finite value if W is compact.

Theorem 9. *Assume that a pair of a statistical model and a prior is a standard form and that λ and m are the real log canonical threshold and its multiplicity. Then the following asymptotic expansion holds for $t \to +0$,*

$$\delta(t - w^{2k})|w^h|b(w)\chi(w) = t^{\lambda-1}(-\log t)^{m-1}D(w)$$
$$+o(t^{\lambda-1}(-\log t)^{m-1}), \qquad (5.28)$$

where $\chi(w)$ and $D(w)$ are defined in eq.(5.24) and (5.27) respectively.

Proof. For an arbitrary C^∞-class function $\psi(w)$, we define

$$v(t) = \int_W \delta(t - w^{2k})|w^h|b(w)\chi(w)\psi(w)dw.$$

The Mellin transform of $v(t)$ is defined by

$$(Mv)(z) = \int_0^\infty v(t)t^z dt,$$

where $z \in \mathbb{C}$. Then by the definition,

$$(Mv)(z) = \int_W w^{2kz}|w^h|b(w)\chi(w)\psi(w)dw.$$

Since W is compact, there exists $D > 0$ such that

$$(Mv)(z) = \int_{[0,D]^d} w^{2kz}|w^h|b(w)\psi(w)dw.$$

By the mean value theorem, there exists w^* such that $|w^*| \leq |w|$ and

$$b(w)\psi(w) = b(0, w_\beta)\psi(0, w_\beta) + \sum_{j=1}^m w_j(\partial/\partial w_j)(b(w^*)\psi(w^*)). \qquad (5.29)$$

A complex function of $z \in \mathbb{C}$ defined by

$$\int_{[0,D]^m} w_\alpha^{2kz}|w_\alpha^h|dw_\alpha = \prod_{j=1}^m \frac{D^{2k_j z + h_j + 1}}{2k_j z + h_j + 1}$$

$$= \frac{1}{(z+\lambda)^m} \prod_{j=1}^m \frac{1 + (D^{2k_j(z+\lambda)} - 1)}{2k_j}$$

has a pole at $z = -\lambda$ with the order m. Therefore by using eq.(5.29),

$$
(Mv)(z) \;=\; \frac{1}{(z+\lambda)^m}\Big(\prod_{j=1}^{m}\frac{1}{2k_j}\Big)\int_{[0,D]^{d-m}} w_\beta^\mu b(0,w_\beta)\psi(0,w_\beta)dw_\beta
$$

$$
+O\Big(\frac{1}{(z+\lambda)^{m-1}}\Big).
$$

Note that the integration of the first term in the right hand side is finite because the redundant index satisfies $\mu_j > -1$. Then by using the inverse Mellin transform shown in Remark 30,

$$
v(t) \;=\; \frac{t^{\lambda-1}(-\log t)^{m-1}}{(m-1)!}\Big(\prod_{j=1}^{m}\frac{1}{2k_j}\Big)\int_{[0,D]^{d-m}} w_\beta^\mu b(0,w_\beta)\psi(0,w_\beta)dw_\beta
$$

$$
+O\Big(t^{\lambda-1}(-\log t)^{m-2}\Big),
$$

which completes Theorem 9. $\square$

Remark 30. The Mellin transform of

$$
y(t) = \begin{cases} t^{\lambda-1}(-\log t)^{m-1} & (0 < t < 1) \\ 0 & (\text{otherwise}) \end{cases}
$$

is given by

$$
(My)(z) = \frac{(m-1)!}{(z+\lambda)^m},
$$

which can be derived by the recursive partial integration.

Remark 31. Theorem 9 shows the following fact. The state density function $\delta(t - w^{2k})|w^h|$ is not a well-defined hyperfunction when $t \to +0$. However, it is asymptotically given by the well-defined hyperfunction $D(w)$,

$$
\delta(t - w^{2k})|w^h|b(w)\chi(w) \cong t^{\lambda-1}(-\log t)^{m-1}D(w),
$$

and its asymptotic behavior as a function of t is determined by the real log canonical threshold λ and its multiplicity m.

Example 35. Let us study a state density function

$$
\delta(t - x^6)|x|^2 b(x)\chi(x),
$$

where $b(x) > 0$. This case corresponds to $k = 3$ and $h = 2$ in the above theorem. Hence $\lambda = (h+1)/(2k) = 1/2$ and $m = 1$. The redundant index is empty.

$$
C(k,m) = 2^m(m-1)!k = 6.
$$

By Theorem 9, as $t \to +0$,

$$\delta(t - x^6)|x^2|b(x)\chi(x) = \frac{t^{-1/2}}{6}\delta(x)b(0) + o(t^{-1/2}).$$

Example 36. Let us study a state density function on

$$\delta(t - x^4 y^6 z^8)|x^1 y^2 z^6|b(x, y, z)\chi(x, y, z),$$

where $b(x, y, z) > 0$. The multi-indexes are $k = (2, 3, 4)$ and $h = (1, 2, 6)$, hence

$$\lambda = \min\left\{\frac{1 + .1}{4}, \frac{2 + 1}{6}, \frac{6 + 1}{8}\right\},$$

which shows $\lambda = 1/2$ and $m = 2$, resulting that

$$C(k, m) = 2^m (m - 1)! k_1 \cdot k_2 = 2^2 \cdot (2 - 1) \cdot 2 \cdot 3 = 24.$$

The redundant multi-index is $\mu_3 = -2 \cdot (1/2) \cdot 4 + 6 = 2$. By Theorem 9, as $t \to 0$,

$$\delta(t - x^4 y^6 z^8)|x^1 y^2 z^6|b(x, y, z)\chi(x, y, z)$$
$$\cong \frac{1}{24} t^{-1/2}(-\log t)\, \delta(x)\delta(y)z^2 b(0, 0, z).$$

Remark 32. Let $K(w)$ and $\varphi(w)$ be an average log density ratio function and a prior, respectively. The zeta function is defined by

$$\zeta(z) = \int K(w)^z \varphi(w) dw \quad (z \in \mathbb{C}),$$

which is equal to the Mellin transform of the state density function. Then $K(w)$ is an analytic function on $\mathrm{Re}(z) > 0$, which can be analytically continued to a meromorphic function whose poles are all real and negative values. Let the largest pole of $\zeta(z)$ be $(-\lambda)$ and its order be m. Then λ and m are equal to the real log canonical threshold and its multiplicity, respectively.

Remark 33. In the above definitions, it is assumed that W is contained in $[0, D]^d$ for some $D > 0$. For the other cases, we can use the same procedures. Let $\sigma = (\sigma_1, \sigma_2, ..., \sigma_d)$, where $\sigma_i = 1$ or $\sigma_i = -1$. The set of all such variables is denoted by Σ_d. For $\sigma \in \Sigma_d$, We define

$$\sigma w = (\sigma_1 w_1, \sigma_2 w_2, ..., \sigma_d w_d).$$

Note that for an arbitrary σ, $\sigma^2 w = w$ and $|\sigma w^k| = |w^k|$. Also we define a set of parameters $\sigma(S)$ by

$$\sigma(S) = \{\sigma w \; ; \; w \in S\}.$$

If $W \subset \sigma([0, D]^d)$, we define a generalized version of $D(w)$ by

$$D_\sigma(w) = \frac{1}{C(k, m)} \, \delta(\sigma w_\alpha) \, |w_\beta^\mu| \, b(w) \chi(\sigma w). \tag{5.30}$$

Then

$$\delta(t - w^{2k})|w^h|b(w)\chi(\sigma w)$$
$$= \delta(t - (\sigma w)^{2k})|(\sigma w)^h|b(w)\chi(\sigma w)$$
$$= t^{\lambda-1}(-\log t)^{m-1}D_\sigma(w) + o(t^{\lambda-1}(-\log t)^{m-1}).$$

Therefore, the asymptotic form of the state density function for the case $\sigma W \subset [0, D]^d$ results in Theorem 9. If $W \subset [-D, D]^d$, the state density function is the sum of the

$$\delta(t - w^{2k})|w^h|b(w)$$
$$= \sum_{\sigma \in \Sigma_d} \delta(t - w^{2k})|w^h|b(w)\chi(\sigma w)$$
$$= t^{\lambda-1}(-\log t)^{m-1}\Big(\sum_{\sigma \in \Sigma_d} D_\sigma(w)\Big) + o(t^{\lambda-1}(-\log t)^{m-1}).$$

By applying Theorem 9, the state density function is represented by the sum of $\sigma \in \Sigma_d$. Note that

$$w^k \, \delta(t - w^{2k})|w^h|b(w)\chi(\sigma w)$$
$$= (\sigma^k) \, t^{\lambda-1/2}(-\log t)^{m-1}D_\sigma(w) + o(t^{\lambda-1/2}(-\log t)^{m-1}).$$

5.3   Asymptotic Free Energy

In this section, we study the asymptotic behavior of the free energy or the minus log marginal likelihood. The normalized posterior function $\Omega(w)$ is defined by

$$\Omega(w) = \frac{\varphi(w) \prod\limits_{i=1}^{n} p(X_i|w)}{\prod\limits_{i=1}^{n} p(X_i|w_0)}. \tag{5.31}$$

This function is in proportion to the posterior density as a function of the parameter w. The posterior density is an exponential function of n, whereas the normalized posterior function is in proportion to $(\log n)^{m-1}/n^\lambda$, where λ and m are the real log canonical threshold and its multiplicity respectively.

Theorem 10. *Assume that a pair of a statistical model and a prior has a standard form and that λ and m are the real log canonical threshold and its multiplicity, respectively. If the set of parameters W is contained in $[0, D]^d$, then*

$$\Omega(w) = \frac{(\log n)^{m-1}}{n^\lambda} D(w) \int_0^\infty dt \, t^{\lambda-1} \exp(-t + \sqrt{t}\,\xi_n(w))$$

$$+ o_p\Big(\frac{(\log n)^{m-1}}{n^\lambda}\Big), \tag{5.32}$$

where $\xi_n(w)$ and $D(w)$ are functions defined by eqs. (5.17) and (5.27), respectively.

Proof. If $w \in W \subset [0, D]^d$, by using eq.(5.22),

$$\begin{aligned}
\Omega(w) &= \exp(-nK_n(w))\varphi(w) \\
&= \exp(-nw^{2k} + \sqrt{n}w^k\xi_n(w))|w^h|b(w) \\
&= \int_0^\infty \delta(t - w^{2k})|w^h| \exp(-nt + \sqrt{nt}\,\xi_n(w))b(w)dt.
\end{aligned}$$

By replacing $t := t/n$ and by Theorem 9, it follows that

$$\begin{aligned}
\Omega(w) &= \int_0^\infty \frac{dt}{n} \delta\Big(\frac{t}{n} - w^{2k}\Big)|w^h| \exp(-t + \sqrt{t}\,\xi_n(w))\, b(w) \\
&= \int_0^\infty \frac{dt}{n} \Big(\frac{t}{n}\Big)^{\lambda-1} \Big(\log\Big(\frac{n}{t}\Big)\Big)^{m-1} D(w) \exp(-t + \sqrt{t}\,\xi_n(w)) \\
&\quad + o_p((\log n)^{m-1}/n^\lambda).
\end{aligned}$$

Since

$$\Big(\log\Big(\frac{n}{t}\Big)\Big)^{m-1} = (\log n)^{m-1} + O((\log n)^{m-2}),$$

we obtain Theorem 10. $\qquad\qquad\qquad\qquad\qquad\qquad\qquad\qquad\square$

Theorem 11. *Assume the same condition as in Theorem 10. Then the free energy or the minus log marginal likelihood*

$$F_n = -\log \int \varphi(w) \prod_{i=1}^n p(X_i|w)dw$$

has the asymptotic expansion

$$\begin{aligned}
F_n &= nL_n(w_0) + \lambda \log n - (m-1)\log\log n \\
&\quad - \log\Big(\int dw D(w) \int_0^\infty dt \, t^{\lambda-1} \exp(-t + \sqrt{t}\,\xi_n(w))\Big) + o_p(1),
\end{aligned}$$

where

$$L_n(w_0) = -\frac{1}{n} \sum_{i=1}^{n} \log p(X_i|w_0).$$

Proof. By the definition of $\Omega(w)$ in eq.(5.31),

$$
\begin{aligned}
F_n &= -\log\Big(\int \Omega(w) \prod_{i=1}^{n} p(X_i|w_0)dw\Big) \\
&= nL_n(w_0) - \log\Big(\int \Omega(w)dw\Big).
\end{aligned}
$$

Combining this equation and Theorem 10 completes the theorem.   □

Remark 34. If W is contained in $[-D, D]^d$, then by Remark 33,

$$
\begin{aligned}
&\delta(t - w^{2k})|w^h|b(w) \\
&= t^{\lambda-1}(-\log t)^{m-1}\Big(\sum_{\sigma\in\Sigma_d} D_\sigma(w)\Big) + o(t^{\lambda-1}(-\log t)^{m-1}).
\end{aligned}
$$

Therefore,

$$
\begin{aligned}
F_n &= nL_n(w_0) + \lambda \log n - (m-1)\log\log n \\
&- \log\Big(\sum_{\sigma\in\Sigma_d} \int dw D_\sigma(w) \int_0^\infty dt\, t^{\lambda-1} \exp(-t + \sqrt{t}\,\sigma^k\,\xi_n(w))\Big) + o_p(1).
\end{aligned}
$$

In other words, the main order terms are same as in Theorem 11, whereas the constant order term is different.

Example 37. For a triple of a statistical model, a true distribution, and a prior given in Example 27, $\lambda = 1$ and $m = 2$. A numerical result of $F_n - nL_n(w_0)$ is shown in Figure 5.4. In general, λ and m depend on a true distribution, hence the Theorem 11 cannot be employed if we do not know the true distribution. However, by using the mathematical property of F_n, we can derive the information criterion WBIC by which F_n can be estimated without information of the true distribution (Chapter 8).

5.4   Renormalized Posterior Distribution

In this section, we derive the asymptotic behaviors of the generalization, training, and cross validation losses and WAIC. By using $\Omega(w)$ in eq.(5.31), the posterior expected value of an arbitrary function $y(w)$ is rewritten as

$$\mathbb{E}_w[y(w)] = \frac{\int y(w)\Omega(w)dw}{\int \Omega(w)dw}. \tag{5.33}$$

This is the definition of the posterior average. However, the behavior for $n \to \infty$ cannot be derived directly from this definition. By using the standard form, the posterior distribution can be represented by a product of a function of n and the fluctuation function. The renormalized posterior distribution is necesary for the fluctuation part.

Definition 19. (Renormalized posterior distribution) Assume the same condition as Theorem 10. For an arbitrary function $z(t, w)$, the renormalized posterior distribution is defined by

$$\langle z(t,w)\rangle = \frac{\int dw D(w) \int_0^\infty dt\, z(t,w)\, t^{\lambda-1}\, \exp(-t + \sqrt{t}\, \xi_n(w))}{\int dw D(w) \int_0^\infty dt\, t^{\lambda-1}\, \exp(-t + \sqrt{t}\, \xi_n(w))}. \tag{5.34}$$

For a given function $f(w)$, the expectation operator $\langle f(w)\rangle$ depends on $\xi_n(w)$. If the function $\xi_n(w)$ must be explicitly represented, a notation

$$\langle f(w)\rangle = \langle f(w)\rangle_{\xi_n} \tag{5.35}$$

is used. Note that $\langle f(w)\rangle$ is a random variable.

Remark 35. By using notation $w = (w_\alpha, w_\beta)$ in Definition 18,

$$D(w) = D(w_a, w_b) \propto \delta(w_a) w_b^\mu,$$

the posterior average can be rewritten as

$$\langle z(t,w)\rangle = \frac{\int dw_\beta\, w_\beta^\mu \int_0^\infty dt\, z(t,0,w_\beta)\, t^{\lambda-1}\, e^{-t+\sqrt{t}\, \xi_n(0,w_\beta)}}{\int dw_\beta\, w_\beta^\mu \int_0^\infty dt\, t^{\lambda-1}\, e^{-t+\sqrt{t}\, \xi_n(0,w_\beta)}}. \tag{5.36}$$

Hence, the renormalized posterior distribution does not depend on the set of values $\{z(t, w_\alpha, w_\beta); |w_\alpha| > 0\}$. The set $\{(0, w_\beta)\}$ is contained in the set of the optimal parameters W_0. In other words, the renormalized posterior distribution is the probability distribution on W_0.

Remark 36. The renormalized posterior distribution is defined for the case when $W \subset [0, D]^d$ for some $D > 0$. For general cases in Remark 33 it is defined by replacing

$$\int dw \, D(w) \mapsto \sum_{\sigma \in \Sigma_d} \int dw D_\sigma(w).$$

The same theorems and lemmas hereafter hold.

Theorem 12. *The renormalized posterior distribution satisfies*

$$\langle t \rangle = \lambda + \frac{1}{2} \langle \sqrt{t} \, \xi_n(w) \rangle. \tag{5.37}$$

Proof. By the definition,

$$\langle t \rangle = \frac{\displaystyle \int dw D(w) \int_0^\infty dt \, t^\lambda \, e^{-t + \sqrt{t} \, \xi_n(u)}}{\displaystyle \int dw D(w) \int_0^\infty dt \, t^{\lambda-1} \, e^{-t + \sqrt{t} \, \xi_n(u)}}. \tag{5.38}$$

By applying the partial integration over t and using $\lambda > 0$,

$$\int_0^\infty e^{-t} t^\lambda e^{\sqrt{t} \xi_n(w)} dt$$

$$= -\left[e^{-t} t^\lambda e^{\sqrt{t} \xi_n(w)} \right]_0^\infty + \int_0^\infty e^{-t} \frac{d}{dt} (t^\lambda e^{\sqrt{t} \xi_n(w)}) dt$$

$$= \lambda \int_0^\infty e^{-t} t^{\lambda-1} e^{\sqrt{t} \xi_n(w)} dt$$

$$+ \int_0^\infty e^{-t} t^\lambda e^{\sqrt{t} \xi_n(w)} \xi_n(w)/(2\sqrt{t}) dt.$$

By applying this equation to eq.(5.38), we obtain Theorem 12. $\qquad \square$

There are asymptotic relations between the posterior distribution and the renormalized one, by which the behaviors of the log density ratio function and its average

$$f(x, w) = \log \frac{p(x|w_0)}{p(x|w)} = w^k a(x, w), \tag{5.39}$$

$$K(w) = \int f(x, w) q(x) dx, \tag{5.40}$$

are clarified.

Theorem 13. (Scaling law) *Assume the same condition as used in Theorem 10. For an arbitrary positive integer s,*

$$\mathbb{E}_w[f(x,w)^s] = \frac{1}{n^{s/2}}\langle\,(\sqrt{t}\,a(x,w))^s\,\rangle + o_p(\frac{1}{n^{s/2}}), \qquad (5.41)$$

$$\mathbb{E}_w[K(w)^s] = \frac{1}{n^s}\langle t^s\,\rangle + o_p(\frac{1}{n^s}). \qquad (5.42)$$

Proof. Let us prove the first half. The latter half can be proved in the same way.

$$\mathbb{E}_w[f(x,w)^s] = \frac{\displaystyle\int w^{sk}a(x,w)^s\Omega(w)dw}{\displaystyle\int \Omega(w)dw}.$$

Let $N(s)$ be the numerator of this equation. Then $N(0)$ is equal to the denominator and

$$N(s) = \int dw\, w^{sk}a(x,w)^s \exp(-nw^{2k}+\sqrt{n}w^k\xi_n(w))w^h b(w)$$

$$= \int_0^\infty dt \int dw\, t^{s/2}a(x,w)^s b(w)\delta(t-nw^{2k})w^h \exp(-t+\sqrt{t}\xi_n(w))$$

$$= \frac{(\log n)^{m-1}}{n^{\lambda+s/2}}\int_0^\infty dt \int dw D(w)a(x,w)^s t^{\lambda+s/2-1}\,\exp(-t+\sqrt{t}\,\xi_n(w))$$

$$+o_p\Big(\frac{(\log n)^{m-1}}{n^{\lambda+s/2}}\Big),$$

where we used eq.(5.32) in Theorem 10. Therefore $N(s)/N(0)$ satisfies the first half. $\qquad\square$

Remark 37. Let $\ell \geq 2$. Assume that a statistical model and a prior have the standard form. Then we can prove

$$\sup_{|\alpha|\leq 1}\Big|\Big(\frac{d}{d\alpha}\Big)^\ell \mathcal{G}_n(\alpha)\Big| = O_p(\frac{1}{n^{\ell/2}})$$

$$\sup_{|\alpha|\leq 1}\Big|\Big(\frac{d}{d\alpha}\Big)^\ell \mathcal{T}_n(\alpha)\Big| = O_p(\frac{1}{n^{\ell/2}})$$

in the same way as we proved Theorem 5. In fact,

$$f(x,w) = w^k a(x,w).$$

Hence

$$|f(x, w)| \leq |w^k||a(x, w)|.$$

By using Lemma 8,

$$
\begin{aligned}
\left|\left(\frac{d}{d\alpha}\right)^\ell \mathcal{G}_n(\alpha)\right| &\leq c_k \mathbb{E}_X\left[\frac{\mathbb{E}_w[|f(X, w)|^\ell \exp(-\alpha f(X, w))]}{\mathbb{E}_w[\exp(-\alpha f(X, w))]}\right] \\
&= c_k \mathbb{E}_X\left[\{\sup_w |a(X, w)|\}\frac{\mathbb{E}_w[|w^k|^\ell \exp(-\alpha f(X, w))]}{\mathbb{E}_w[\exp(-\alpha f(X, w))]}\right] \\
&= O_p(1/n^{\ell/2}).
\end{aligned}
$$

In the same way,

$$
\begin{aligned}
\left|\left(\frac{d}{d\alpha}\right)^\ell \mathcal{T}_n(\alpha)\right| &\leq c_k \frac{1}{n}\sum_{i=1}^n \frac{\mathbb{E}_w[|f(X_i, w)|^\ell \exp(-\alpha f(X_i, w))]}{\mathbb{E}_w[\exp(-\alpha f(X_i, w))]} \\
&= O_p(1/n^{\ell/2}).
\end{aligned}
$$

Therefore, we can apply the basic Theorem 3 also to this case.

Definition 20. By using the renormalized posterior distribution, the fluctuation of the renormalized posterior distribution $\text{Fluc}(\xi_n)$ is defined by

$$\text{Fluc}(\xi_n) = \mathbb{E}_X[\langle ta(X, w)^2\rangle - \langle \sqrt{t}a(X, w)\rangle^2].$$

Here $\text{Fluc}(\xi_n)$ is a functional of ξ_n, because the expected value using the renormalized posterior distribution $\langle\ \rangle$ depends on ξ_n.

Theorem 14. *Assume the same condition as in Theorem 10. Then*

$$G_n = L(w_0) + \frac{1}{n}\left(\lambda + \frac{1}{2}\langle\sqrt{t}\xi_n(w)\rangle - \frac{1}{2}\text{Fluc}(\xi_n)\right) + o_p(\frac{1}{n}), \quad (5.43)$$

$$T_n = L_n(w_0) + \frac{1}{n}\left(\lambda - \frac{1}{2}\langle\sqrt{t}\xi_n(w)\rangle - \frac{1}{2}\text{Fluc}(\xi_n)\right) + o_p(\frac{1}{n}), \quad (5.44)$$

$$C_n = L_n(w_0) + \frac{1}{n}\left(\lambda - \frac{1}{2}\langle\sqrt{t}\xi_n(w)\rangle + \frac{1}{2}\text{Fluc}(\xi_n)\right) + o_p(\frac{1}{n}), \quad (5.45)$$

$$W_n = L_n(w_0) + \frac{1}{n}\left(\lambda - \frac{1}{2}\langle\sqrt{t}\xi_n(w)\rangle + \frac{1}{2}\text{Fluc}(\xi_n)\right) + o_p(\frac{1}{n}), \quad (5.46)$$

Proof. By the Theorem 3 and the above lemma, the generalization, cross validation, and training losses are obtained by calculating $\mathcal{G}_n'(0)$, $\mathcal{G}_n''(0)$, $\mathcal{T}_n'(0)$, and $\mathcal{T}_n''(0)$, by applying the scaling law. Using eq.(5.37),

$$
\begin{aligned}
\mathcal{G}_n'(0) &= L(w_0) + \mathbb{E}_w[K(w)] \\
&= L(w_0) + \frac{1}{n}\langle t\rangle + o_p(1/n) \\
&= L(w_0) + \frac{1}{n}\left(\lambda + \frac{1}{2}\langle\sqrt{t}\xi_n(w)\rangle\right) + o_p(1/n),
\end{aligned}
$$

and

$$
\begin{aligned}
\mathcal{G}_n''(0) &= \mathbb{E}_X[\mathbb{E}_w[f(X,w)^2] - \mathbb{E}_w[f(X,w)]^2] \\
&= \frac{1}{n}\mathbb{E}_X[\langle ta(X,w)^2\rangle - \langle \sqrt{t}a(X,w)\rangle^2] + o_p(1/n) \\
&= \frac{1}{n}\mathrm{Fluc}(\xi_n) + o_p(1/n).
\end{aligned}
$$

Hence we obtained eq.(5.43).

$$
\begin{aligned}
\mathcal{T}_n'(0) &= L_n(w_0) + \mathbb{E}_w[K_n(w)] \\
&= L_n(w_0) + \frac{1}{n}\langle t - \sqrt{t}\xi_n(w)\rangle + o_p(1/n) \\
&= L_n(w_0) + \frac{1}{n}\left(\lambda - \langle\frac{1}{2}\sqrt{t}\xi_n(w)\rangle\right) + o_p(1/n),
\end{aligned}
$$

and

$$
\begin{aligned}
\mathcal{T}_n''(0) &= \frac{1}{n}\sum_{i=1}^{n}\left\{\mathbb{E}_w[f(X_i,w)^2] - \mathbb{E}_w[f(X_i,w)]^2\right\} \\
&= \frac{1}{n^2}\sum_{i=1}^{n}\left\{\langle ta(X_i,w)^2\rangle - \langle\sqrt{t}a(X_i,w)\rangle^2\right\} + o_p(1/n).
\end{aligned}
$$

By using the law of large numbers for a functional case,

$$
\sup_{w,v}\left|\frac{1}{n}\sum_{i=1}^{n}a(X_i,w)a(X_i,v) - \mathbb{E}_X[a(X,w)a(X,v)]\right| = o_p(1),
$$

the difference between $n\mathcal{G}_n''(0)$ and $n\mathcal{T}_n''(0)$ goes to zero in probability when $n \to \infty$. Thus eq.(5.44) is obtained. □

Remark 38. The standard deviations of T_n, C_n, and W_n are $O_p(1/\sqrt{n})$, because the standard deviation of $L_n(w_0)$ is $O_p(1/\sqrt{n})$. The averages and standard deviations of the four random variables $G_n - L(w_0)$, $T_n - L_n(w_0)$, $C_n - L_n(w_0)$, and $W_n - L_n(w_0)$ are $O_p(1/n)$. They are asymptotically given by the linear combination of two random variables $\langle\sqrt{t}\xi_n(w)\rangle$ and $\mathrm{Fluc}(\xi_n)$. It should be emphasized that Theorem 14 holds even if the true distribution is not realizable by and singular for a statistical model.

Remark 39. By Theorem 14, the convergences in probability hold,

$$
\begin{aligned}
n(G_n - L(w_0)) + n(C_n - L_n(w_0)) &\to 2\lambda, \\
n(G_n - L(w_0)) + n(W_n - L_n(w_0)) &\to 2\lambda,
\end{aligned}
$$

where λ is the real log canonical threshold. That is to say, for a given triple $(q(x), p(x|w), \varphi(w))$, if $(C_n - L_n(w_0))$ is smaller then $(G_n - L_n(w_0))$ is larger. W_n has the same property as C_n. This is the generalized version of Remark 25 in regular theory. These properties are very important when we employ the cross validation loss and WAIC in statistical model selection and hyperparameter optimization. If a sample consists of independent random variables, then it automatically follows that $\mathbb{E}[C_n] = \mathbb{E}[G_{n-1}]$ by the definition of the cross validation loss. However in order to derive the variance of the cross validation error $C_n - L_n(w_0)$, we need mathematical theory.

Lemma 22. *Let $\xi(w)$ be the Gaussian process which is uniquely characterized by eq.(5.20) and eq.(5.21). Then*

$$\mathbb{E}_\xi[\langle \sqrt{t}\xi(w)\rangle] = \mathbb{E}_\xi[\mathrm{Fluc}(\xi)].$$

Proof. Since $\mathbb{E}[G_{n-1}] = \mathbb{E}[C_n]$, by the above theorem,

$$\mathbb{E}_{\xi_n}[\langle \sqrt{t}\xi_n(w)\rangle] = \mathbb{E}_{\xi_n}[\mathrm{Fluc}(\xi_n)] + o(1).$$

As $n \to \infty$, $\xi_n(u) \to \xi(u)$ in distribution, which completes the lemma. $\square$

Remark 40. Lemma 22 is shown by the relation between expected values of generalization and cross validation losses, based on the assumption that $X_1, X_2, ..., X_n$ are independent. However, the condition of independent random variables is not necessary for Lemma 22. In fact, it holds in some cases when $X_1, X_2, ..., X_n$ are not independent. See the next section.

Definition 21. The constant

$$2\nu = \mathbb{E}_\xi[\langle \sqrt{t}\xi(w)\rangle] = \mathbb{E}_\xi[\mathrm{Fluc}(\xi)]$$

is called the singular fluctuation.

Theorem 15. *Let λ and ν be the real log canonical threshold and the singular fluctuation. Then the averages of the generalization loss, the cross validation loss, the training loss, and WAIC are asymptotically given by*

$$\mathbb{E}[G_n] = L(w_0) + \frac{\lambda}{n} + o(\frac{1}{n}), \tag{5.47}$$

$$\mathbb{E}[T_n] = L(w_0) + \frac{1}{n}(\lambda - 2\nu) + o(\frac{1}{n}), \tag{5.48}$$

$$\mathbb{E}[C_n] = L(w_0) + \frac{\lambda}{n} + o(\frac{1}{n}), \tag{5.49}$$

$$\mathbb{E}[W_n] = L(w_0) + \frac{\lambda}{n} + o(\frac{1}{n}). \tag{5.50}$$

Proof. By Theorem 14 and Lemma 22, we obtain Theorem 15. $\square$

Definition 22. The functional variance is defined by

$$V_n = \frac{1}{n} \sum_{i=1}^{n} \Big\{ \mathbb{E}_w[(\log p(X_i|w)^2] - \mathbb{E}_w[\log p(X_i|w)]^2 \Big\}.$$

Lemma 23. *When* $n \to \infty$, $n\mathbb{E}[V_n] \to 2\nu$.

Proof. Since $\log p(X_i|w_0)$ is a constant function of a parameter w, the random variable V_n can be represented by

$$V_n = \frac{1}{n} \sum_{i=1}^{n} \Big\{ \mathbb{E}_w[f(X_i, w)^2] - \mathbb{E}_w[f(X_i, w)]^2 \Big\}.$$

The asymptotic equivalence of V_n and $\mathrm{Fluc}(\xi_n)$ was shown in the part $\mathcal{T}_n''(0)$ in Theorem 14. $\square$

Theorem 16. *¿Equations of states in Bayesian statistics¿ Assume that a statistical model has a standard form.*

$$\mathbb{E}[G_n] = \mathbb{E}\Big[C_n\Big] + o(\frac{1}{n}) \tag{5.51}$$

$$\mathbb{E}[G_n] = \mathbb{E}\Big[W_n\Big] + o(\frac{1}{n}). \tag{5.52}$$

These equations hold even if a true distribution is singular for or unrealizable by a statistical model.

Proof. This theorem is immediately derived from Theorem 15. $\square$

Remark 41. If the true distribution is regular for a statistical model, then the higher order equivalence can be derived. See Chapter 8.

Remark 42. Both C_n and W_n are estimators of G_n. In typical statistical inferences, the difference of them is very small. If $X_1, X_2, ..., X_n$ are independent, then both eq.(5.51) and eq.(5.52) hold. If $X_1, X_2, ..., X_n$ are dependent, then it is not ensured that eq.(5.51) hold. Under some conditions such as conditonal independence, eq.(5.52) holds. Therefore, even if $X_1, X_2, ..., X_n$ are dependent, if eq.(5.52) holds and C_n is asymptotically equivalent to W_n, then also eq.(5.51) holds. See the next section.

5.5   Conditionally Independent Case

In this book, we mainly study a case when a sample X^n consists of random variables which are independently subject to the same probability distribution. However, there are several important cases when such an assumption does not hold. In this section, we study a conditionally independent case. In this situation, the differences between the cross validation and information criteria are clarified.

In this section, we study a case when

$$x^n = (x_1, x_2, ..., x_n)$$

is fixed or may be dependent. The random variables $Y^n = (Y_1, Y_2, ..., Y_n)$ are independently subject to a true conditional distribution

$$\prod_{i=1}^{n} q(y_i|x_i).$$

Then $(Y_1, Y_2, ..., Y_n)$ are independent, but $((x_1, Y_1), (x_2, Y_2), ..., (x_n, Y_n))$ are dependent.

Remark 43. The conditionally independent condition allows the following cases.

1. The set $(x_1, x_2, ..., x_n)$ consists of fixed points.

2. The set $(x_1, x_2, ..., x_n)$ is a time sequence.

3. The set $(x_1, x_2, ..., x_n)$ is not independent.

Note that, in such cases, prediction and estimation are different from each other. In general, prediction for a new point x_{n+1} has no meaning, whereas the estimation of $q(y|x_i)$ using $p(y|x_i, w)$ is well-defined.

We define a statistical model and a prior by

$$p(y_i|x_i, w), \quad \varphi(w).$$

For a given sample (x^n, Y^n), a posterior density is defined by

$$p(w|x^n, Y^n) = \frac{1}{Z_n} \varphi(w) \prod_{i=1}^{n} p(Y_i|x_i, w),$$

where Z_n is a partition function or the marginal likelihood,

$$Z_n = \int \varphi(w) \prod_{i=1}^{n} p(Y_i|x_i, w)dw,$$

which can be understood as the estimated probability density function of Y^n. Also we define a Bayesian estimation of the conditional distribution by

$$p(y|x_i, x^n, Y^n) = \mathbb{E}_w[p(y|x_i, w)].$$

This predictive distribution is not a prediction for a new point but an estimation of Y at the trained point x_i. A prediction for a new point x can be defined by the same equation,

$$p(y|x, x^n, Y^n) = \mathbb{E}_w[p(y|x, w)],$$

however, its generalization loss cannot be defined because there does not exist a probability distribution of x. The generalization loss, the cross validation loss, the training loss, and WAIC are defined by

$$G_n = -\frac{1}{n}\sum_{i=1}^{n}\int q(y|x_i)\log p(y|x_i, x^n, Y^n)dy,$$

$$T_n = -\frac{1}{n}\sum_{i=1}^{n}\log p(Y_i|x_i, x^n, Y^n),$$

$$C_n = -\frac{1}{n}\sum_{i=1}^{n}\log p(Y_i|x_i, x^n \setminus x_i, Y^n \setminus Y_i),$$

$$W_n = T_n + \frac{1}{n}\sum_{i=1}^{n}\mathbb{V}_w[\log p(Y_i|x_i, w)].$$

Also in this case we can construct Bayesian theory for both regular and standard cases. For example, eq.(3.20), eq.(3.21), and eq.(3.23) in Theorem 3 hold, resulting that eq.(5.43), eq(5.44), and eq.(5.46) in Theorem 14 hold. However, there are several different points from the independent case. Because x^n is not subject to the same probability distribution, the average generalization loss is not equal to the cross validation loss,

$$\mathbb{E}[G_{n-1}] \neq \mathbb{E}[C_n].$$

In other words, if x^n is dependent, then we lost the relation between the generalization loss and the cross validation loss. Moreover, the asymptotic expansion of C_n in eq.(3.22) in Theorem 3 does not hold in general,

$$C_n = \mathcal{T}_n(-1) \neq -\mathcal{T}'_n(0) + \frac{1}{2}\mathcal{T}''_n(0) + o_p(\frac{1}{n}).$$

Consequently, eq.(5.45) in Theorem 14 does not hold.

$$C_n \neq L_n(w_0) + \frac{1}{n}\Big(\lambda - \frac{1}{2}\langle\sqrt{t}\xi_n(w)\rangle + \frac{1}{2}\mathrm{Fluc}(\xi_n)\Big) + o_p(\frac{1}{n}).$$

Remark 44. (Numerical problem of importance samping cross validation) In conditionally independent cases, the cross validation loss has the other problem. Even in a conditionally independent case, the cross validaiton loss C_n satisfies

$$C_n = \frac{1}{n}\sum_{i=1}^{n}\log\mathbb{E}_w[1/p(Y_i|x_i,w)],$$

which can be numerically calculated by using posterior parameters, hence it is called the importance sampling cross validation. However, if a leverage sample point (x_i, Y_i) is contained, in other words, if $p(Y_i|x_i,w)$ is very small for a posterior parameter w, then the posterior average $\mathbb{E}_w[1/p(Y_i|x_i,w)]$ diverges. Therefore eq.(3.22) in Theorem 3 does not hold in general.

Since $\mathbb{E}[G_{n-1}] \neq \mathbb{E}[C_n]$, Lemma 22 cannot be derived via the cross validation loss. However, we can prove the following Lemma 24, by which eq.(5.47), eq.(5.48), and eq.(5.49) in Theorem 15 and eq.(5.52) in Theorem 16 hold. However, neither eq.(5.49) in Theorem 15 nor eq.(5.51) in Theorem 16 holds. In other words, the cross validation loss cannot be employed in conditionally independent cases whereas WAIC can be.

Let us show theoretical strcutures of the conditionally independent cases. Let W_0 be the set of parameters which minimize

$$L(w) = -\frac{1}{n}\sum_{i=1}^{n}\int q(y|x_i)\log p(y|x_i,w)dy,$$

and w_0 be a parameter contained in W_0. Two functions $f(x_i, y, w)$ and $K(w)$ are defined by

$$f(x_i, y, w) = \log\frac{p(y|x_i, w_0)}{p(y|x_i, w)},$$

$$K(w) = \frac{1}{n}\sum_{i=1}^{n}\int q(y|x_i)f(x_i, y, w)dy.$$

If a set of a statistical model and a prior is a standard form, there exist $a(x_i, y, w)$ and $b(w)$ such that

$$
\begin{aligned}
f(x_i, y, w) &= w^k a(x_i, y, w), \\
K(w) &= w^{2k}, \\
\varphi(w) &= |w^h| \, b(w),
\end{aligned}
$$

where both $k > 0$ and $h \geq 0$ are multi-indexes and $b(w) > 0$ in a neighborhood of the origin. A stochastic process $\xi_n(w)$ is defined by

$$
\xi_n(w) = \frac{1}{\sqrt{n}} \sum_{i=1}^{n} \{ w^k - a(x_i, Y_i, w) \}.
$$

Note that $\xi_n(w)$ is an empirical process composed of dependent random variables in general. It follows that

$$
\begin{aligned}
\mathbb{E}[\xi_n(w)] &= 0, \\
\mathbb{E}[\xi_n(w)\xi_n(u)] &= \frac{1}{n} \sum_{i=1}^{n} \mathbb{E}_Y[a(x_i, Y, w)a(x_i, Y, u)] - w^k u^k.
\end{aligned}
$$

A Gaussian process $\xi(w)$ which satisfies the following conditions is uniquely determined.

$$
\begin{aligned}
\mathbb{E}_\xi[\xi(w)] &= 0, \\
\mathbb{E}_\xi[\xi(w)\xi(u)] &= \frac{1}{n} \sum_{i=1}^{n} \mathbb{E}_Y[a(x_i, Y, w)a(x_i, Y, u)] - w^k u^k.
\end{aligned}
$$

Then $\langle \sqrt{t}\xi_n(w) \rangle$, $\langle \sqrt{t}\xi(w) \rangle$, $\mathrm{Fluc}(\xi_n)$, and $\mathrm{Fluc}(\xi)$ are defined in the same way. We can derive the following lemma without using the cross validation loss.

Lemma 24. *The same statement as Lemma 22 holds,*

$$
\mathbb{E}_\xi[\langle \sqrt{t}\xi(w) \rangle] = \mathbb{E}_\xi[\mathrm{Fluc}(\xi)].
$$

Proof. In this proof we use a notation $\mathbb{E}_\xi[\ ] = \mathbb{E}[\ ]$. Let us use a decomposition of the Gaussian process,

$$
\xi(w) = \sum_{j=1}^{\infty} g_j \xi_j(w),
$$

where $\{g_j\}$ is a set of random variables which are independently subject to the standard normal distribution $\mathcal{N}(0,1)$. If $K(g(w)) = 0$ then $w^k = 0$ and

$$\mathbb{E}[\xi(w)\xi(v)] = \frac{1}{n}\sum_{i=1}^{n}\mathbb{E}_Y[a(x_i, Y, w)a(x_i, Y, u)] = \sum_{j=1}^{\infty}\xi_j(w)\xi_j(v). \quad (5.53)$$

A Gaussian process satisfies

$$\mathbb{E}[g_j F(g_j)] = \mathbb{E}\Big[\frac{\partial}{\partial g_j}F(g_j)\Big]$$

for an arbitrary integrable function $F(\ )$. Let S be the integration operator defined by

$$S[\ ] = \int dw D(w) \int_0^{\infty} dt\, t^{\lambda-1}\exp(-t)[\ ].$$

Then

$$\mathbb{E}[\langle\sqrt{t}\xi(w)\rangle] = \mathbb{E}\Big[\frac{S[\sqrt{t}\xi\exp(\sqrt{t}\xi)]}{S[\exp(\sqrt{t}\xi)]}\Big]$$

$$= \sum_{j=1}^{\infty}\mathbb{E}\Big[\frac{\partial}{\partial g_j}\Big(\frac{S[\sqrt{t}\xi_j\exp(\sqrt{t}\xi)]}{S[\exp(\sqrt{t}\xi)]}\Big)\Big]$$

$$= \sum_{j=1}^{\infty}\Big\{\mathbb{E}\Big[\frac{S[t\xi_j^2\exp(\sqrt{t}\xi)]}{S[\exp(\sqrt{t}\xi)]}\Big] - \mathbb{E}\Big[\frac{S[\sqrt{t}\xi_j\exp(\sqrt{t}\xi)]}{S[\exp(\sqrt{t}\xi)]}\Big]^2\Big\}.$$

The last equation is equal to $\mathrm{Fluct}(\xi)$, which completes the lemma. $\square$

Remark 45. For a finite n, if the convergence in distribution $\xi_n(w) \to \xi(w)$ holds, then

$$\mathbb{E}_{\xi_n}[\langle\sqrt{t}\xi_n(w)\rangle] = \mathbb{E}_{\xi_n}[\mathrm{Fluc}(\xi_n)] + o(1).$$

Therefore, the above equation also holds asymptotically. In the proof of Lemma 24, the partial integration over the functional space is effectively employed. Therefore, we understand that in independent cases, the cross validation is mathematically equivalent to the partial integration over the functional space.

Example 38. Let us illustrate an example in which the convergence $\xi_n(w) \to \xi(w)$ holds in a conditionally independent case. Let $y \in \mathbb{R}$, $x \in \mathbb{R}^N$, and $w \in \mathbb{R}^d$. We study a case when a statistical model $p(y|x, w)$ is given by

$$p(y|x, w, s) = \sqrt{\frac{s}{2\pi}}\exp(-(s/2)(y - F(x, w))^2),$$

where $F(x, w)$ is a function from $\mathbb{R}^N \times \mathbb{R}^d$ to $\mathbb{R}$. Assume that x^n is a set of fixed points and Y^n is taken from $p(y|x_i, w_0, s_0)$ for some true w_0 and s_0. Then

$$
\begin{aligned}
f(x_i, y, w, s) &= (s/2)(y - F(x_i, w))^2 + (1/2) \log s \\
&\quad -(s_0/2)(y - F(x_i, w_0))^2 + (1/2) \log s_0,
\end{aligned}
$$

and

$$
K(w, s) = \frac{1}{n} \sum_{i=1}^{n} \int p(y|x_i, w_0, s_0) f(x_i, y, w, s) dy.
$$

Hence if a statistical model is a standard form,

$$
\begin{aligned}
w^k \xi_n(w, s) &= \frac{s_0 - s}{2\sqrt{n}} \sum_{i=1}^{n} \{(Y_i - F(x_i, w_0))^2 - 1/s\} \\
&\quad + \frac{s}{\sqrt{n}} \sum_{i=1}^{n} (Y_i - F(x_i, w))(F(x_i, w) - F(x_i, w_0)),
\end{aligned}
$$

which converges to a Gaussian process as $n \to \infty$, because $\{Y_i - F(x_i, w_0)\}$ is a set of independent random variables which are subject to the same probability distribution.

Example 39. (Influential observation) A statistical model and a prior are defined by

$$
\begin{aligned}
p(y|x, a, s) &= \sqrt{\frac{s}{2\pi}} \exp(-(s/2)(y - ax)^2), \\
\varphi(a, s) &\propto s \exp(-(s/2)(\rho + \mu a^2)),
\end{aligned}
$$

where hyperparameters are set as $\mu = \rho = 0.01$. The true conditional density is defined by $q(y|x) = p(y|x, w_0, s_0)$, where $w_0 = 0.2$ and $s_0 = 100$. The set of inputs is given by

$$
\begin{aligned}
x_i &= 0.1i \quad (i = 1, 2, ..., n - 1), \\
x_n &= R,
\end{aligned}
$$

where $n = 10$ and Y^n are independently taken from $q(y|x_i)$. The inputs of data $0 < x_i < 1$ for $i = 1, 2, .., n - 1$, whereas the last input x_n is a leverage sample point because it is set as $R = 1, 2, 3, 4, 5$. Figure 5.2 shows the generalization, the cross validation, and WAIC errors for given R,

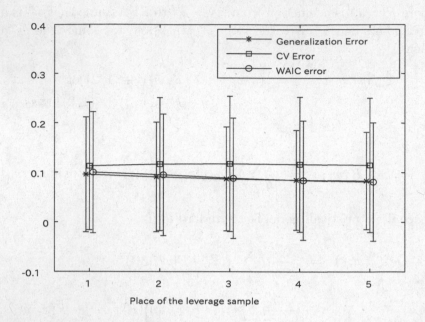

Figure 5.2: Influential observation. The generalization, cross validation, and WAIC errors are compared for the case when a leverage sample point is contained. The horizontal line shows the place of the leverage sample point whereas other sample points are in the interval $[0, 1]$. If the leverage sample point is far from others, then the variance of the cross validation error becomes large.

respectively. Here the generalization error is measured not by the average of any probability distribution but by the emprical average of x^n. As the value R is larger, the effect of the leverage sample point becomes larger, resulting that the average of the cross validation error becomes different from that of the generalization error, and the variance becomes larger.

Example 40. (High dimensional case) Let us study a statistical model of $y \in \mathbb{R}$ for a given $x \in \mathbb{R}^d$ with a parameter (a, s) $(a \in \mathbb{R}^d, s > 0)$,

$$p(y|x, a, s) = \sqrt{\frac{s}{2\pi}} \exp(-\frac{s}{2}(y - a \cdot x)^2)$$

and a prior

$$\varphi(a, s) \propto s^r \exp(-\frac{s}{2}(\rho + \mu\|a\|^2)),$$

where $r = d/2$ and $\rho = 0.005$, and $\mu = 0.005$ are hyperparameters. In this example, we studied cases when $d = 20$, or 50, and $n = 100$. A true density was set as $q(y|x) = p(y|x, a_0, s_0)$, where $a_0 = (0.5, 0.5, ..., 0.5)$ and $s_0 = 100$. Then the posterior density of (s, a) is given by

$$p(s|X^n, Y^n) \propto s^{n/2 + r - d/2} \exp(-Cs/2),$$
$$p(a|s, X^n, Y^n) = \mathcal{N}_d(B, A^{-1}/s),$$

where $\mathcal{N}_d(b, S)$ is the d dimensional normal distribution whose average is b and covariance matrix is S and

$$A = \sum_{i=1}^{n} X_i X_i^T + \mu,$$

$$B = A^{-1}\left(\sum_{i=1}^{n} X_i Y_i,\right)$$

$$C = -\text{tr}(ABB^T) + \sum_{i=1}^{n} Y_i^2 + \rho.$$

Therefore by using the simple Monte Carlo method, we approximated the generalization loss G_n, the cross validation loss C_n, and WAIC W_n.

(1) $d = 20$, $n = 100$. Firstly, let X^n consist of independent random variables which are subject to the normal distribution $\mathcal{N}_d(0, I)$ where I is the identity matrix. Then G_n is mesured by the average over this distribution. The experimental averages and standard deviations for 1000 independent trials were

$$C_n - S_n = 0.130, \quad 0.034,$$
$$W_n - S_n = 0.123, \quad 0.033,$$
$$G_n - S = 0.129, \quad 0.030,$$

and the estimated errors were

$$\mathbb{E}[\,|G_n - S - (C_n - S_n)|\,] = 0.0462,$$
$$\mathbb{E}[\,|G_n - S - (W_n - S_n)|\,] = 0.0456.$$

(2) $d = 20$, $n = 100$. Secondly, X^n was generated from $\mathcal{N}_d(0, I)$ and then fixed, and G_n was defined by empirical mean over the fixed X^n. The experimental averages and standard deviations of $C_n - S_n$ and $W_n - S_n$ for 1000 independent trials were same as (1), whereas those of $G_n - S$ were

$$G_n - S = 0.108, \quad 0.024$$

and the estimated errors were

$$\mathbb{E}[\,|G_n - S - (C_n - S_n)|\,] = 0.047,$$
$$\mathbb{E}[\,|G_n - S - (W_n - S_n)|\,] = 0.044.$$

(3) $d = 40$, $n = 100$. The experimental averages and standard deviations for 1000 independent trials were

$$G_n - S = 0.425, \quad 0.030,$$
$$C_n - S_n = 0.428, \quad 0.047,$$
$$W_n - S_n = 0.402, \quad 0.047,$$

and the estimated errors were

$$\mathbb{E}[\,|G_n - S - (C_n - S_n)|\,] = 0.056,$$
$$\mathbb{E}[\,|G_n - S - (W_n - S_n)|\,] = 0.058.$$

(4) $d = 40$, $n = 100$. Secondly, X^n was generated from $\mathcal{N}_d(0, I)$ and then fixed. The experimental averages and standard deviations of $C_n - S_n$ and $W_n - S_n$ for 1000 independent trials were same as (3), whereas those of $G_n - S$ were

$$G_n - S = 0.336, \quad 0.021,$$

and the estimated errors were

$$\mathbb{E}[\,|G_n - S - (C_n - S_n)|\,] = 0.098,$$
$$\mathbb{E}[\,|G_n - S - (W_n - S_n)|\,] = 0.078.$$

If d/n is not so large, $W_n - S_n$ is the better estimator of $G_n - S$ than $C_n - S_n$, for both indepedent and fixed x^n. If d/n is larger, then statistical estimation is not accurate. If x^n is independent, $C_n - S_n$ is the better estimator of $G_n - S$ than $W_n - S_n$. If x^n is fixed, $W_n - S_n$ is the better estimator of $G_n - S$ than $C_n - S_n$. Our recommendation in practical problems is as follows. If either the cross validation loss or WAIC can be calculated by the Markov Chain Monte Carlo method, then the other can also be calculated by almost the same computational time. Therefore, we recommend that both of them would be calculated and compared. If they are different, x^n may be dependent or contains the leverage sample point.

Example 41. (Time sequence) In time series analysis, we sometimes use a statistical model

$$Y_i = aY_{i-1} + bY_{i-2} + cY_{i-3} + \text{noise}.$$

This model is represented by

$$p(Y_i|Y_{i-1}, Y_{i-2}, Y_{i-3}, a, b, c),$$

which is equivalent to

$$p(Y_i|x_i, a, b, c),$$

where

$$x_i = (Y_{i-1}, Y_{i-2}, Y_{i-3}) \in \mathbb{R}^3.$$

The sample point x_i depends on Y_j $(j \neq i)$. However, this model is equivalent to the model in which $\{Y_i\}$ are independent under the condition that x^n are given. If we adopt the assumption that a true probability distribution satisfies the same conditional independence as the statistical model, WAIC can be applied to evaluation of estimating accuracy at the set of empirical points x^n. Moreover, if the cross validation loss has almost the same value as WAIC, then the cross validation loss also can be employed as an approximated value of WAIC.

5.6 Problems

1. Let k be a positive integer and $\lambda = 1/k$. Prove the following equations.

$$\int_0^1 \exp(-nx^k)dx = \frac{\lambda}{n^\lambda}\left(\int_0^n \exp(-y)\, y^{\lambda-1}\, dy\right),$$

$$\int_0^1 \delta(t - x^k)dx = \lambda\, t^{\lambda-1},$$

$$\int_0^1 (x^k)^z dx = \frac{\lambda}{z + \lambda},$$

where $n > 0$, $0 < t < 1$, and $\text{Re}(z) > -\lambda$.

2. Let k be a positive integer and $\lambda = 1/k$. Assume that $\varphi(x)$ is a C_1 class

function on $(-\varepsilon, 1]$, where $\varepsilon > 0$. Then prove that

$$\int_0^1 \exp(-nx^k)\varphi(x)dx = \frac{\lambda\Gamma(\lambda)}{n^\lambda}\varphi(0) + o(1/n^\lambda),$$

$$\int_0^1 \delta(t - x^k)\varphi(x)dx = \lambda\, t^{\lambda-1}\varphi(0) + o(t^{\lambda-1}),$$

$$\int_0^1 (x^k)^z\varphi(x)dx = \frac{\lambda}{z+\lambda}\varphi(0) + g(z),$$

as $n \to \infty$, $t \to +0$, and $g(z)$ is an analytic function on $\mathrm{Re}(z) > -\lambda$ which can be analytically continued to an analytic function on $\mathrm{Re}(z) > -2\lambda$.

3. Let $f(x, y)$ and $g(x, y)$ be polynomials of (x, y) which satisfy $f(0, 0) = 0$. Let U be an open set which contains $(0, 0)$ and $U^* = U \backslash \{(x, y); f(x, y) = 0\}$.
(1) Make an example of a set $f(x, y)$ and $g(x, y)$ which satisfies

$$\sup_{(x,y)\in U^*} \left| \frac{g(x, y)}{f(x, y)} \right| < 1,$$

when $g(x, y)/f(x, y)$ is not a polynomial.
(2) Assume that $f(x, y) = x^2 y^2$. Prove that if

$$\sup_{(x,y)\in U^*} \left| \frac{g(x, y)}{f(x, y)} \right| < 1,$$

then $g(x, y)/f(x, y)$ is a polynomial.
Explain the mathematical difference between (1) and (2).

4. A neural network is defined by a conditional density of $x, y \in \mathbb{R}$,

$$p(y|x, a, b) = \frac{1}{\sqrt{2\pi}} \exp(-\frac{1}{2}(y - a\tanh(bx))^2).$$

We study a case when a probability density of x is the uniform distribution of $[-2, 2]$. A true conditional density is set as $p(y|x, 0, 0)$, and the uniform prior for (a, b) on $[-1, 1] \times [-1, 1]$ is adopted. Then the true distribution is realizable by and singular for a statistical model. Figure 5.3 shows the posterior distributions for six independent sets of (X^n, Y^n), where $n = 100$. Even if the true parameter $(a_0, b_0) = (0.1, 0.1)$, the posterior distributions have almost the same shapes as the case $(a_0, b_0) = (0, 0)$. In statistical model selection and hypothesis testing, we often have to determine $(a_0, b_0) = (0, 0)$

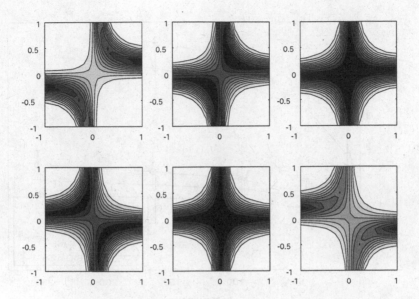

Figure 5.3: Fluctuation of posterior distribution in neural network. Posterior distributions of six independent sets of X^n are shown for the case $n = 100$. The set of the true parameters is $\{(a, b); ab = 0\}$.

or $(a_0, b_0) \neq (0, 0)$ based on a sample. Discuss whether we can apply regular statistical theory to neural networks or not.

5. The statistical model, the true density, and the prior are set as in the above neural network. Hence the posterior distribution cannot be approximated by any normal distribution. Note that λ and m, which depend on the true distribution, are $\lambda = 1/2$ and $m = 2$. Let $(\hat{a}, \hat{b})$ be the maximum *a posteriori* estimator. Since the uniform prior of (a, b) on $[-1, 1] \times [-1, 1]$ is adopted, it is equal to the maximum likelihood estimator in this case. The log loss function is

$$L_n(a, b) = -\frac{1}{n} \sum_{i=1}^{n} \log p(Y_i | X_i, a, b).$$

The free energy F_n is equal to

$$F_n = -\log \int \prod_{i=1}^{n} \exp(-n L_n(w)) \varphi(a, b) da db.$$

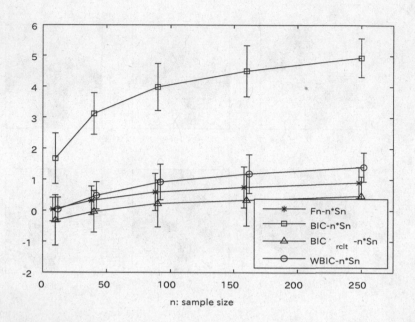

Figure 5.4: Free energy and its estimators in neural network. Free energy, BIC, BIC using RLCT, and WBIC are compared in a simple neural network. BIC is larger than the free energy. Both BIC using RLCT and WBIC can approximate the free energy.

In general, it is difficult to calculate the integration over the parameter set in F_n. However, in this case, it can be approximiated by the Riemann sum because the dimension of the parameter space is small (2). Its estimators, BIC, BIC using RLCT, and WBIC are given by

$$\text{BIC} = nL_n(\hat{a}, \hat{b}) + (d/2)\log n,$$
$$\text{BIC}_{\text{rclt}} = nL_n(\hat{a}, \hat{b}) + \lambda \log n - (m-1)\log\log n,$$
$$\text{WBIC} = \mathbb{E}_{(a,b)}^{(\beta)}[nL_n(a, b)],$$

where $d = 2$ and $\mathbb{E}_{(a,b)}^{(\beta)}[\ \ ]$ shows the expectation value by the posterior distribution with the inverse temperature $\beta = 1/\log n$. The empirical entropy S_n is equal to $L_n(0, 0)$. Figure 5.4 shows $F_n - nS_n$, $\text{BIC} - Sn$, $\text{BIC}_{rclt} - Sn$, and $\text{WBIC} - Sn$. Both BIC using RLCT and WBIC can approximate the free energy, whereas BIC not. Discuss the difference of BIC, BIC using RLCT, and WBIC from the viewpoint of estimators of the free energy for $n \to \infty$.

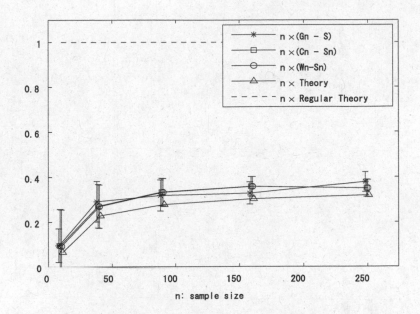

Figure 5.5: Generalization error and its estimators in neural network. The generalization error, the cross validation error, the WAIC error, the theoretical value, and the regular theoretical value are compared. The generalization error can be approximated by the cross validation and WAIC, but not by the regular theory.

6. The statistical model, the true density, and the prior are set as in the above neural network. Figure 5.5 shows $n(G_n - S)$, $n(C_n - S_n)$, $n(W_n - Sn)$, the theoretical value by standard theory, and that by regular theory. Regular theory cannot be employed in this case. Since the generalization error $(G_n - S)$ and the cross validation error $(C_n - S_n)$ have the asymptotically inverse correlation,

$$n(G_n - S) + n(C_n - S_n) = \frac{2\lambda}{n} + o_p(1/n),$$

where $\lambda = 1/2$ in this case. Hence it needs many sets of (X^n, Y^n) to numerically show

$$\mathbb{E}[G_n - S] = \mathbb{E}[C_n - S_n] + o(1/n).$$

In this experiment, the results by 1000 independent trials are shown. Discuss the difference of standard theory and regular theory for $n \to \infty$.

Chapter 6

General Posterior Distribution

In the previous chapter, we introduced a standard form of a statistical model and a prior, based on which mathematical laws of the free energy and the generalization loss were proved. In this chapter, we explain that many models and priors can be made into standard forms by using the algebraic geometric transform. Then the posterior distribution is represented as a finite mixture of locally standard forms,

$$p(w) = \sum \text{Standard form}.$$

As a result, the same theorems of the previous chapter also hold in many statistical models and priors. Also we show the difference between the Bayesian and the maximum *a posteriori* methods. This chapter consists of the following sections.

(1) In Bayesian estimation, the set of parameters can be understood as a union of local parameter sets.

(2) Resolution theorem in algebraic geometry makes an arbitrary statistical model a locally standard form.

(3) General theory of Bayesian statistics is established.

(4) The generalization losses of the maximum likelihood and *a posteriori* methods are derived.

6.1 Bayesian Decomposition

In Bayesian statistics, the posterior distribution can be decomposed as a sum of local distributions. Let $(p(x|w), \varphi(w))$ be a pair of a statistical model and

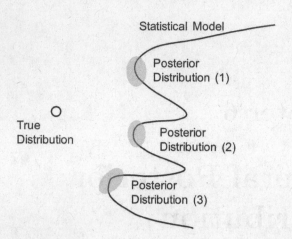

Figure 6.1: Division of parameter set. Bayesian posterior distribution can be understood as the mixture of several distributions.

a prior. A decomposition of a prior $\varphi(w)$ is

$$\varphi(w) = \sum_j \varphi_j(w),$$

where $\varphi_j(w) \geq 0$. Then the partition function or the marginal likelihood is given by

$$Z_n = \sum_j \int \varphi_j(w) \prod_{i=1}^n p(X_i|w)dw.$$

The function $X^n \mapsto Z_n$ is a probability density of X^n according to the pair of the statistical model $p(x|w)$ and a prior $\varphi(w)$, hence

$$Z_n(j) = \int \varphi_j(w) \prod_{i=1}^n p(X_i|w)dw$$

defines a probability distribution on the pairs $\{p(x|w), \varphi_j(w)\}$. The posterior average of an arbitrary function $f(w)$ is also given by

$$\mathbb{E}_w[f(w)] = \frac{\displaystyle\sum_j \int f(w)\varphi_j(w) \prod_{i=1}^n p(X_i|w)dw}{\displaystyle\sum_j \int \varphi_j(w) \prod_{i=1}^n p(X_i|w)dw}.$$

Therefore Bayesian estimation can be studied from the viewpoint of the local parameter sets.

The support of $\varphi_j(w)$ is defined as the closure of nonzero set of $\varphi_j(w)$,

$$\text{supp } \varphi_j = \overline{\{w \in W; \varphi_j(w) > 0\}}.$$

Let W_0 be the set of all parameters that attain the minimum of the average log loss function $L(w)$,

$$W_0 = \{w \in W; L(w) = \min_{w'} L(w)\}.$$

If $\{\text{supp } \varphi_j\} \cap W_0$ is the empty set, then $Z_n(j)$ converges to zero faster than the others. If $\{\text{supp } \varphi_j\} \cap W_0$ is not the empty set, and if a statistical model and a prior have a standard form in each local subset, then by using the local real log canonical threshold λ_j and its multiplicty m_j, the local partition function can be rewritten as

$$Z_n(j) = c_j \frac{(\log n)^{m_j - 1}}{n^{\lambda_j}} \prod_{i=1}^{n} p(X_i | w_0),$$

where c_j is a constant order random variable and $w_0 \in W_0$. Let us define

$$\lambda = \min\{\lambda_j; j\},$$
$$m = \max \#\{m_j; \lambda_j = \lambda\},$$

where $\#\{m_j; \lambda_j = \lambda\}$ is the maximum number of j that attains $\lambda = \lambda_j$. Then asymptotically

$$Z_n \approx \left(\sum_j c_j\right) \frac{(\log n)^{m-1}}{n^{\lambda}} \prod_{i=1}^{n} p(X_i | w_0),$$

where $\sum_j$ is the summation over $\{j; \lambda = \lambda_j, m_j = m\}$, resulting that the free enegy or the minus log marginal likelihood has the same asymptotic form as the theorems of previous chapters.

Example 42. In Figure 6.1, the circle shows the true distribution and the curved line is a set of statistical models. The posterior distribution is made of local parameters (1), (2), and (3). In such a case, the parameter set can be divided into three parts, and the posterior distribution can be represented by their summation.

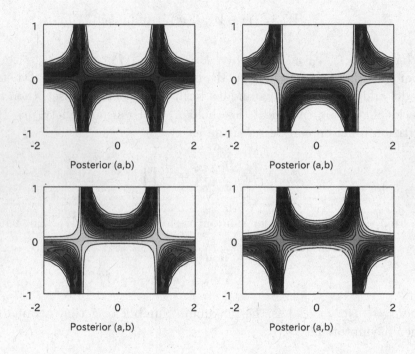

Figure 6.2: Artificial examples of posterior distributions. The set of true parameters is $b(a-1)(a+1) = 0$. The posterior distribution can be understood as a mixture of local standard forms.

Example 43. Let a statistical model and a prior be

$$p(y|x,a,b) = \frac{1}{\sqrt{2\pi}}\exp(-\frac{1}{2}\{y - b\tanh((a-1)x)\tanh((a+1)x)\}^2),$$

$$\varphi(a,b) = \begin{cases} 1/8 & (|a| \leq 2, |b| \leq 1) \\ 0 & (\text{otherwise} \end{cases},$$

where the set of all parameters is

$$W = \{(a,b); |a| \leq 2, |b| \leq 1\}.$$

Also let (X^n, Y^n) be a set of random variables which are independently taken from a true distribution $q(x)p(y|x,0,0)$, where $q(x)$ is the uniform distribution on $[-2,2]$. Then the set of true parameters is $\{(a,b); b(a - 1)(a+1) = 0\}$. Figure 6.2 shows four posterior distributions for different independent sets (X^n, Y^n). By dividing parameter set

$$W = \{(a,b) \in W; a \leq 0\} \cup \{(a,b) \in W; a \geq 0\},$$

the posterior distribution is represented as a mixture of standard forms. To each distribution, we can apply the theory in the previous chapter because $b(a-1)$ and $b(a+1)$ can be made standard form by $a_1 = a-1$ and $a_2 = a+1$.

6.2 Resolution of Singularities

Even if the posterior distribution cannot be approximated by any normal distribution, there exists division of parameter set such that the average log density ratio function can be normal crossing in each local parameter set. The resolution theorem in algebraic geometry is the mathematical base for statistical analysis of general Bayesian statistics.

Theorem 17 (Hironaka theorem). *Let W be a compact which is the closure of an open set in $\mathbb{R}^d$. Assume that $K(w) \geq 0$ is a nonzero analytic function on W and that the set $\{w \in W; K(w) = 0\}$ is not empty. Then there exist $\epsilon > 0$, $\{W_\ell; W_\ell \subset W\}$, and $\{U_\ell; U_\ell \subset \mathbb{R}^d\}$ which satisfy*

$$\{w \in W \; ; \; K(w) \leq \epsilon\} = \bigcup_\ell W_\ell,$$

and, in each pair W_ℓ and U_ℓ, there exists an analytic map $g : U_\ell \to W_\ell$ which satisfies

$$K(g(u)) = u^{2k}, \tag{6.1}$$

$$|g'(u)| = b(u)|u^h|, \tag{6.2}$$

where $|g'(u)|$ is the absolute value of the determinant of the Jacobian matrix of $w = g(u)$,

$$|g'(u)| = \left|\det\left(\frac{\partial w_i}{\partial u_j}\right)\right|,$$

and k, h ($k > 0$, $h \geq 0$) are d-dimensional multi-indexes, and $b(u) > 0$.

Example 44. Let a parameter set be $[-0.5, 0.5]^2$. A statistical model and a prior are defined by

$$p(y|x, a, b) = \frac{1}{(2\pi\sigma^2)^{1/2}} \exp(-\frac{1}{2\sigma^2}(y - (a^4 - a^2 b + b^3)x)^2),$$

$$\varphi(a, b) = 1,$$

where $\sigma = 0.01$. Let a true distribution be $q(y|x) = p(y|x, 0, 0)$ and $n = 100$. The average log density ratio function is

$$K(a, b) = \frac{1}{2\sigma^2}(a^4 - a^2 b + b^3)^2.$$

Hence the set of true parameters is

$$a^4 - a^2 b + b^3 = 0.$$

We define a blowing-up by

$$a = a',$$
$$b = 3a'b'.$$

It follows that

$$a^4 - a^2 b + b^3 = a'^3(a' - 3b' + 27b'^3).$$

Figure 6.3 shows the set of true parameters on (a, b) plane, its blown-up on (a', b') plane, the posterior distribution of (a, b) and the blown-up posterior distribution of (a', b'). Note that the determinant of Jacobian matrix of this blowing-up is

$$|g'(a', b')| = 3|a'|,$$

hence the blown-up posterior distribution is defined using this equation. On the (a, b) plane, the origin is a not normal crossing singularity, whereas, on the (a', b') plane, $(0, 0)$, $(0, 1/3)$ and $(0, -1/3)$ are normal crossing singularities. Hence on the (a', b') plane, the posterior distribution is a mixture of standard forms.

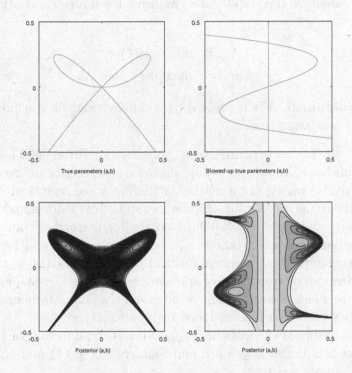

Figure 6.3: Example of resolution theorem. The set of true parameters on (a, b), its blown-up on (a', b'), the posterior distribution of (a, b) and the blown-up posterior distribution of (a', b'). By using the resolution of singularities, any singularity can be understood as an image of normal crossing singularities. By using resoluion theorem, most statistical models can become standard forms, to which the theorems in the previous chapter can be applied.

Definition 23. In the resolution theorem, $g(u)$, k, h, and $b(u)$ depend on U_ℓ, although such dependence is not explicitly represented because of simple description. If such dependence is necessary for description, then representations $g_\ell(u)$, k_ℓ, h_ℓ, and $b_\ell(u)$ are used. The real log canonical threshold λ_ℓ and its multiplicity m_ℓ are determined in each U_ℓ. Then the real log canonical threshold and its multiplicity of $K(w)$ and $\varphi(w)$ are defined by

$$\lambda = \min\{\lambda_\ell \,;\, \ell\},$$
$$m = \max\{m_\ell \,;\, \lambda_\ell = \lambda\}.$$

By the definition, λ is a positive real number and m is a positive integer which is not larger than d.

Remark 46. Let us explain several points about Hironaka Theorem.
(1) In this book, knowledge about algebraic geometry is not necessary. However, a reader who is not a mathematician may see an introductive book to nonmathematicians [82], in whicn many statistical models and learning machines are studied. A book [69] is written for mathematicians who are not majoring in algebraic geometry. A book [51] is basic and famous in algebraic geometry. From a computational point of view, [14] is recommended for students. The resolution of singularities is proved by [37] and studied as one of the main themes in algebraic geometry [42]. Mathematical relation to algebraic analysis is introduced by [9] and [41].
(2) The number of elements of $\{W_\ell\}$, which is equal to that of $\{U_\ell\}$, is finite.
(3) Since $k > 0$ and $h \geq 0$ are multi-indices, if eq(6.1) and eq.(6.2) can be more explicitly written,

$$K(g(u)) = (u_1)^{2k_1}(u_2)^{2k_2}\cdots(u_d)^{2k_d},$$
$$|g'(u)| = b(u)|(u_1)^{h_1}(u_2)^{h_2}\cdots(u_d)^{h_d}|.$$

Note that $|g'(u)| = 0$ if and only if $u^h = 0$, the map $w = g(u)$ is one-to-one if and only if $|u^h| > 0$. Such a function $w = g(u)$ is called a birational map.
(4) In this book, manifold theory is not required, but if a reader already studied manifold theory, then $\cup_\ell W_\ell$ and $\cup_\ell U_\ell$ are compact subsets of manifolds and $g(u) = \{g_\ell(u)\}$ is a map from a manifold to a manifold. In other words, $g_\ell(u)$ can be understood as a restriction to a local coordinate of a map from a manifold to another manifold.
(5) In general, for a given function $K(w)$, neither $\{U_\ell\}$ nor $w = g(u)$ is unique. Neither k nor h is unique. However, the real log canonical threshold λ and its multiplicity m are uniquely determined and do not depend

on the choice of $w = g(u)$. Such values are called *birational invariants*. The real log canonical threshold is an important birational invariant in high dimensional algebraic geometry, whereas it is also important in Bayesian statistics. All Bayesian observables are automatically birational invariants, because they do not depend on the choice of parameter representations.

(6) If $K(g(u)) = K_0(u)u^{2k}$ for some $K_0(u) > 0$ and $k_1 > 0$, then by

$$
\begin{aligned}
u'_1 &= K_0(u)^{1/2k_1}u_1, \\
u'_2 &= u_2, \\
&\cdots \\
u'_d &= u_d,
\end{aligned}
$$

eq(6.1) and eq.(6.2) are satisfied about u', because $\epsilon > 0$ can be sufficiently small.

(7) This theorem is called Hironaka's resolution of singularities, which is the fundamental theorem in algebraic geometry. It holds for an arbitrary analytic function $K(w)$. Note that this theorem can be employed even if one does not know the definition of singularities. If $K(w)$ is an analytic function, then it holds even if $K(w) = 0$ does not contain singularities.

(8) Even if $K(w)$ is not an analytic function, there are several cases to which this theorem can be applied. For example, if $K(w)$ is a piecewise analytic function, then this theorem can be applied to each parameter set. If $K(w) = K_0(w)K_1(w)$, where $K_0(w) > 0$ may not be analytic and $K_1(w)$ is analytic, then the resolution theorem can be applied to $K_1(w)$ and the same statistical theory can be derived.

(9) If a prior has a hyperparameter, then $\{\lambda_\ell\}$ and $\{m_\ell\}$ are functions of the hyperparameter. In general they may be discontinuous or nondifferentiable, which is the main reason for phase transition of the posterior distribution. See Section 9.4.

(10) For a given function $K(w)$, the Hironaka theorem gives the algebraic algorithm by which the resolution map can be found. However, in general, it is not easy to find the resolution map $w = g(u)$. For several statistical models, the complete resolution maps were found. For others, partial resolution maps were found by which the upper bounds of the real log canonical thresholds were derived. Even if the complete resolution map cannot be founded, its existence enables us to prove universal formula in Bayesian statistics. For example, we have methods by which the generalization loss and the minus log marginal likelihood are estimated without information about the resolution map.

Example 45. (Resolution by projective space) Let $x = (y, z), w = (s, t) \in \mathbb{R}^2$ and

$$W = \{w = (s, t) ; 0 \le s, t \le 1\}.$$

A statistical model and a prior are

$$p(y, z|s; t) = \frac{1}{\pi} \exp(-(y - s)^2 - (z - t)^2),$$
$$\varphi(s, t) = 1.$$

Assume that $q(x) = p(x|0, 0)$. Then the optimal parameter that minimizes the average log loss function is $w_0 = (0, 0)$. The log density ratio function and its average are respectively equal to

$$f(x, w) = s^2 + t^2 - 2ys - 2zt,$$
$$K(w) = s^2 + t^2.$$

This is not a standard form. In Example 26, we showed it can be made a standard form by using a polar coordinate system (r, θ). Here we give another transform.

$$W = W_1 \cup W_2,$$
$$W_1 = \{(s, t) \in W ; s \ge t\},$$
$$W_2 = \{(s, t) \in W ; s \le t\}.$$

Then we prepare two other parameter sets,

$$U_1 = \{(s_1, t_1) ; 0 \le s_1, t_1 \le 1\},$$
$$U_2 = \{(s_2, t_2) ; 0 \le s_2, t_2 \le 1\}.$$

Then by using two maps,

$$s = s_1 = s_2 t_2,$$
$$t = s_1 t_1 = t_2.$$

The log density ratio function and its average are respectively given by

$$f(y, z, s, t) = s_1(s_1 + s_1 t_1^2 - 2y - 2z t_1),$$
$$K(s, t) = s_1^2(1 + t_1^2),$$

and

$$f(y, z, s, t) = t_2(s_2^2 t_2 + t_2 - 2y s_2 - 2z),$$
$$K(s, t) = t_2^2(s_2^2 + 1),$$

both of which are standard forms. The set $U_1 \cup U_2$ is called a projective space. The integration of an arbitrary function $f(s,t)$ can be divided,

$$
\begin{aligned}
\int_W f(s,t)\, ds\, dt &= \int_{W_1} f(s,t)\, ds\, dt + \int_{W_2} f(s,t)\, ds\, dt \\
&= \int_{U_1} f(s_1, s_1 t_1)\, t_1\, ds_1\, dt_1 \\
&\quad + \int_{U_2} f(s_2, s_2 t_2)\, s_2\, ds_2\, dt_2,
\end{aligned}
$$

where we used the absolute values of the Jacobian determinant are t_1 and s_2 respectively. Therefore, we can apply the standard theory to this case, and the real log canonical threshold and its multiplicity are $\lambda = 1$ and $m = 1$.

Remark 47. (1) By the same method as Example 45, a regular posterior distribution can be made standard. Hence the regular statistical theory which requires the positive definiteness of the Fisher information matrix is a very special case of general statistics. In general theory, the Fisher information matrix may contain the eigenvalue zero and the transform from the parameter set to another parameter set may not be diffeomorphism. It is well known that Bayesian statistics gives the better estimation than the maximum likelihood one in the case when the Fisher information matrix is degenerate. Such a fact can be mathematically proved by using algebraic geometry.

(2) If a statistical model has a relatively finite variance, there exist $c_0 > 0$ such that

$$
\int q(x) f(x,w)^2 dx \leq c_0 \int a(x) f(x,w) dx = c_0 K(w).
$$

Hence if both $f(x,w)$ and $K(w)$ are analytic functions of w and if

$$
K(w) = w^{2k},
$$

then there exists a function $a(x,w)$ such that

$$
f(x,w) = w^k\, a(x,w).
$$

Therefore, the form of the log density ratio function is automatically derived from its average function.

(3) If $\varphi(w) = 0$ on the set $\{w; K(w) = 0\}$, then by applying resolution theorem to $K(w)\varphi(w)$, there exists a function $w = g(u)$ such that

$$
\begin{aligned}
K(g(u)) &= u^{2k}, \\
\varphi(g(u))|g'(u)| &= |u^h| b(u),
\end{aligned}
$$

where $b(u) > 0$. Such a method is called simultaneous resolution of singularities in algebraic geometry. Algebraic geometry is known as one of the most abstract mathematics, however, its concrete version gives the essential base of Bayesian statistics.

Example 46. Let $(x, y) \in \mathbb{R}^2$, $w = (a, b, c) \in \mathbb{R}^3$, and

$$W = \{w = (a, b, c) ; |a|, |b|, |c| \leq 1\}.$$

A statistical model and a prior are

$$q(x) = \begin{cases} 1/2 & (|x| \leq 1) \\ 0 & (|x| > 1) \end{cases}, \tag{6.3}$$

$$p(x, y|a, b, c) = \frac{q(x)}{\sqrt{2\pi}} \exp(-\frac{1}{2}(y - aS(bx) - cx)^2), \tag{6.4}$$

$$\varphi(a, b, c) = 1/8, \tag{6.5}$$

where $S(x) = x + x^2$. If a true distribution is $q(x, y|0, 0, 0)$, then the log density ratio function and its average are

$$f(x, y|a, b, c) = \frac{1}{2}\{(aS(bx) + cx)^2 - 2y(a(S(bx) + cx)\},$$

$$K(a, b, c) = \frac{1}{2}(ab + c)^2 + \frac{1}{6}a^2b^4.$$

Let us divide the parameter sets by

$$\begin{aligned} W_1 &= \{|a| \leq |c|\}, \\ W_2 &= \{|a| \geq |c|, |ab| \leq |ab + c|\}, \\ W_3 &= \{|a| \geq |c|, |ab + c| \leq |ab^2|\}, \\ W_4 &= \{|a| \geq |c|, |ab^2| \leq |ab + c| \leq |ab|\}. \end{aligned}$$

Then $W = \cup_j W_j$. The function $w = g(u)$ is defined by

$$\begin{aligned} \dot{a} &= a_1c_1, & b &= b_1, & c &= c_1, \\ a &= a_2, & b &= b_2c_2, & c &= a_2(1 - b_2)c_2, \\ a &= a_3, & b &= b_3, & c &= a_3b_3(b_3c_3 - 1), \\ a &= a_4, & b &= b_4c_4, & c &= a_4b_4c_4(c_4 - 1). \end{aligned}$$

The corresponding $U_j = \{(a_j, b_j, c_j)\}$ $(j = 1, 2, 3, 4)$ are defined by

$$U_j = \overline{g^{-1}((W_j)^o)},$$

where $(W_j)^o$ is the maximum open set that is contained in W_j and $\overline{g^{-1}((W_j)^o)}$ is the minimum closed set that contains $g^{-1}((W_j)^o)$. Note that the function $w = g(u)$ is one-to-one on $(W_j)^o$, hence g^{-1} is well-defined on such a set. The average log density ratio function is given by

$$
\begin{aligned}
K(a,b,c) &= c_1^2\Big(\frac{1}{2}(a_1 b_1 + 1)^2 + \frac{1}{6}a_1^2 b_1^4\Big), \\
&= a_2^2 c_2^2\Big(\frac{1}{2} + \frac{1}{6}b_2^2 c_2^2\Big), \\
&= a_3^2 b_3^4\Big(\frac{1}{2}c_3^2 + \frac{1}{6}\Big), \\
&= a_4^2 b_4^2 c_4^4\Big(\frac{1}{2} + \frac{1}{6}b_4^2\Big).
\end{aligned}
$$

Hence $K(a,b,c)$ is normally crossing in each local coordinate. The absolute value of the determinant of the Jacobian matrix is

$$
\begin{aligned}
|g'(u)| &= |c_1| \\
&= |a_2 c_2| \\
&= |a_3 b_3^2| \\
&= |a_4 b_4 c_4|^2.
\end{aligned}
$$

The real log canonical threshold λ_j and its multiplicity m_j for each U_j are

$$
\begin{aligned}
\lambda_1 &= 1, & m_1 &= 1, \\
\lambda_2 &= 1, & m_2 &= 2, \\
\lambda_3 &= 3/4, & m_3 &= 1, \\
\lambda_4 &= 3/4, & m_4 &= 1.
\end{aligned}
$$

Hence the real log canonical threshold and its multiplicity of $(K(w), \varphi(w))$ are $\lambda = 3/4$ and $m = 1$ respectively.

Remark 48. For a given average log density ratio function and a prior, there exists a recursive and algebraic algorithm by which an analytic map $w = g(u)$ is found in a finite procedures.

Example 47. By using the example above, we investigate the phase transition of Bayesian statistics. Instead of the prior given in eq.(6.5), let us study the prior which has a hyperparameter $\alpha > 0$,

$$
\varphi(a,b,c|\alpha) = \frac{\alpha}{8}|a|^{\alpha-1}.
$$

Then in each U_j,

$$
\begin{aligned}
\varphi(g(u)) &= (\alpha/8)|(a_1 c_1)^{\alpha-1}| \\
&= (\alpha/8)|a_2^{\alpha-1}| \\
&= (\alpha/8)|a_3^{\alpha-1}| \\
&= (\alpha/8)|a_4^{\alpha-1}|.
\end{aligned}
$$

Therefore,

$$
\begin{aligned}
\varphi(g(u))|g'(u)| &= (\alpha/8)|a_1^{\alpha-1} c_1^{\alpha}| \\
&= (\alpha/8)|a_2^{\alpha} c_2| \\
&= (\alpha/8)|a_3^{\alpha} b_3^2| \\
&= (\alpha/8)|a_4^{\alpha} b_4^2 c_4^2|.
\end{aligned}
$$

The real log canonical threshold λ_j for each U_j is

$$
\begin{aligned}
\lambda_1 &= (\alpha+1)/2, \\
\lambda_2 &= \min\{(\alpha+1)/2, 1\}, \\
\lambda_3 &= \min\{(\alpha+1)/2, 3/4\}, \\
\lambda_4 &= \min\{(\alpha+1)/2, 3/4\}.
\end{aligned}
$$

If $\alpha \neq 1/2$ and $\alpha \neq 1$, then $m_j = 1$. Therefore

$$
\lambda = \begin{cases} (\alpha+1)/2 & (\alpha \leq 1/2) \\ 3/4 & (\alpha > 1/2) \end{cases}.
$$

If $\alpha = 1/2$, then $m = 2$, otherwise $m = 1$. Note that λ is a continuous function of α, but not differentiable at $\alpha = 1/2$. The posterior distributions for $\alpha > 1/2$ and $\alpha < 1/2$ are quite different from each other. The hyperparameter $\alpha = 1/2$ is called the critical point of the phase transition. Because the real log canonical threshold determines the asymptotic properties of the free energy and the generalization loss, hyperparameter control is important for nonregular statistical models.

6.3   General Asymptotic Theory

By using the resolution theorem, in the parameter set $\{w \in W \; ; \; K(w) \leq \epsilon\}$, the pair of a statistical model and a prior can be made a standard form. Here

$\epsilon > 0$ is a sufficiently small constant in resolution theorem. The normalized partition function is represented by the sum

$$Z_n^{(0)} = Z_n^{(1)} + Z_n^{(2)}. \tag{6.6}$$

By using the constant $\epsilon > 0$,

$$Z_n^{(1)} = \int_{K(w) < \epsilon} \exp(-nK_n(w))\varphi(w)dw,$$

$$Z_n^{(2)} = \int_{K(w) \geq \epsilon} \exp(-nK_n(w))\varphi(w)dw.$$

Here $Z_n^{(1)}$ and $Z_n^{(2)}$ are the integrations of parameters in a neighborhood of the optimal parameter set W_0 and the outside respectively.

The nonessential part $Z_n^{(2)}$ can be bounded by the following procedures. By the same definition used in Section 4.1,

$$\gamma_n(w) = \frac{1}{\sqrt{n}} \sum_{i=1}^{n} \frac{K(w) - f(X_i, w)}{\sqrt{K(w)}}$$

and

$$\gamma_n = \sup_{w \in W_0} |\gamma_n(w)|,$$

the following inequality is derived,

$$\begin{aligned} nK_n(w) &= nK(w) - \sqrt{nK(w)}\gamma_n(w), \\ &\geq nK(w)/2 - \gamma_n^2/2. \end{aligned}$$

Therefore

$$\begin{aligned} Z_n^{(2)} &\leq \exp(\gamma_n^2/2) \int_{K(w) \geq \epsilon} \exp(-nK(w)/2)\varphi(w)dw \\ &\leq \exp(-n\epsilon/2 + \gamma_n^2/2). \end{aligned}$$

Hence $Z_n^{(2)} = o_p(\exp(-n\epsilon/3))$. Let us study the essential part $Z_n^{(1)}$. By the resolution theorem, there exists $w = g(u)$ such that

$$\{w \; ; \; K(w) < \epsilon\} \subset \bigcup_j g(U_j),$$

where, in each U_j,

$$\begin{aligned} K(g(u)) &= u^{2k_j}, \\ \varphi(g(u)) &= |u^{h_j}|b_j(u), \end{aligned}$$

where multi-indexes k_j and h_j, and a function $b_j(u)$ depend on U_j. In each U_j, the local real log canonical threshold λ_j, its multiplicity m_j, and local redundant index μ_j are determined by the same method used in the previous chapter. Then $D_j(u)$ is defined by the same manner as eq.(5.27),

$$D_j(u) = \frac{1}{C(k_j, m_j)} \delta(u_\alpha) |u_\beta^{\mu_j}| b_j(u) \chi(u).$$

The normalized posterior function on U_j is

$$\begin{aligned} \Omega(g(u)) &= \frac{\varphi(g(u))|g'(u)| \prod\limits_{i=1}^{n} p(X_i|g(u))}{\prod\limits_{i=1}^{n} p(X_i|w_0)} \\ &= \exp(-nK_n(g(u)))b(u)|u^h| \\ &= \frac{(\log n)^{m_j-1}}{n^{\lambda_j}} D_j(u) \int_0^\infty dt\, t^{\lambda_j-1}\, \exp(-t + \sqrt{t}\, \xi_n(u)) \\ &\quad + o_p\Big(\frac{(\log n)^{m_j-1}}{n^{\lambda_j}}\Big). \end{aligned}$$

This equation shows the asymptotic form of the posterior distribution. When n tends to infinity, only the sets $\{U_j\}$ that maximize

$$\frac{(\log n)^{m_j-1}}{n^{\lambda_j}}$$

affect the asymptotic form, and are called essential local coordinates. In other words, the ratio of $\Omega(g(u))$ of a set $\{U_j\}$ whose $(\log n)^{m_j-1}/n^{\lambda_j}$ is smaller than the essential local coordinate goes to zero in probability. By the Definition 23, a set U_j is an essential local coordinate if and only if $\lambda_j = \lambda$ and $m_j = m$. Let ELC be the set of all suffixes j such that U_j is an essential local coordinates. Then the general theory can be established by the same procedure as the standard theory with the replacement,

$$D(w) \mapsto \sum_{j \in ELC} D_j(u).$$

Let us summarize the general asymptotic theory. The asymptotic free energy is

$$
\begin{aligned}
F_n \;=\; & nL_n(w_0) + \lambda \log n - (m-1)\log\log n \\
& - \log\Big(\sum_{j\in ELC} \int duD_j(u) \int_0^\infty dt\; t^{\lambda-1}\,\exp(-t+\sqrt{t}\,\xi_n(u))\Big) \\
& + o_p(1).
\end{aligned}
$$

The renormalized posterior distribution is defined for an arbitrary function $z(t,u)$,

$$
\langle z(t,u)\rangle = \frac{\displaystyle\sum_{j\in ELC}\int dwD_j(u)\int_0^\infty dt\; z(t,u)\, t^{\lambda-1}\,\exp(-t+\sqrt{t}\,\xi_n(u))}{\displaystyle\sum_{j\in ELC}\int duD_j(u)\int_0^\infty dt\; t^{\lambda-1}\,\exp(-t+\sqrt{t}\,\xi_n(u))}.
$$

Then, the renormalized posterior distribution satisfies

$$
\langle t\rangle = \lambda + \frac{1}{2}\langle \sqrt{t}\,\xi_n(u)\rangle.
$$

The scaling law which connects the original posterior of w and the renormalized one of u is given for an arbitrary positive integer s,

$$
\begin{aligned}
\mathbb{E}_w[f(x,w)^s] &= \frac{1}{n^{s/2}}\langle\, (\sqrt{t}\,a(x,u))^s\,\rangle + o_p\Big(\frac{1}{n^{s/2}}\Big), \\
\mathbb{E}_w[K(w)^s] &= \frac{1}{n^s}\langle t^s\,\rangle + o_p\Big(\frac{1}{n^s}\Big).
\end{aligned}
$$

By using the renormalized posterior distribution, $\mathrm{Fluc}(\xi_n)$ is defined by

$$
\mathrm{Fluc}(\xi_n) \;=\; \mathbb{E}_X[\langle ta(X,u)^2\rangle - \langle\sqrt{t}a(X,u)\rangle^2].
$$

Then the singular fluctuation is defined by

$$
2\nu = \mathbb{E}_\xi[\langle\sqrt{t}\xi(w)\rangle] = \mathbb{E}_\xi[\mathrm{Fluc}(\xi)].
$$

Asymptotic behaviors of the generalization loss, training loss, cross validation loss, and WAIC are given by

$$
G_n = L(w_0) + \frac{1}{n}\Big(\lambda + \frac{1}{2}\langle\sqrt{t}\xi_n(u)\rangle - \frac{1}{2}\mathrm{Fluc}(\xi_n)\Big) + o_p\Big(\frac{1}{n}\Big), \quad (6.7)
$$

$$
T_n = L_n(w_0) + \frac{1}{n}\Big(\lambda - \frac{1}{2}\langle\sqrt{t}\xi_n(u)\rangle - \frac{1}{2}\mathrm{Fluc}(\xi_n)\Big) + o_p\Big(\frac{1}{n}\Big), \quad (6.8)
$$

$$
C_n = L_n(w_0) + \frac{1}{n}\Big(\lambda - \frac{1}{2}\langle\sqrt{t}\xi_n(u)\rangle + \frac{1}{2}\mathrm{Fluc}(\xi_n)\Big) + o_p\Big(\frac{1}{n}\Big), \quad (6.9)
$$

$$
W_n = L_n(w_0) + \frac{1}{n}\Big(\lambda - \frac{1}{2}\langle\sqrt{t}\xi_n(u)\rangle + \frac{1}{2}\mathrm{Fluc}(\xi_n)\Big) + o_p\Big(\frac{1}{n}\Big). \quad (6.10)
$$

respectively, whose expectations are

$$
\begin{aligned}
\mathbb{E}[G_n] &= L(w_0) + \frac{\lambda}{n} + o(\frac{1}{n}), \\
\mathbb{E}[T_n] &= L(w_0) + \frac{1}{n}(\lambda - 2\nu) + o(\frac{1}{n}), \\
\mathbb{E}[C_n] &= L(w_0) + \frac{\lambda}{n} + o(\frac{1}{n}), \\
\mathbb{E}[W_n] &= L(w_0) + \frac{\lambda}{n} + o(\frac{1}{n}),
\end{aligned}
$$

respectively. Note that the convergence in probability holds,

$$
\begin{aligned}
n(G_n - L(w_0)) + n(C_n - L_n(w_0)) &\to 2\lambda, \\
n(G_n - L(w_0)) + n(W_n - L_n(w_0)) &\to 2\lambda.
\end{aligned}
$$

In other words, the generalization error and the cross validation and WAIC errors have the inverse correlation. The functional variance is defined by

$$
V_n = \frac{1}{n} \sum_{i=1}^{n} \Big\{ \mathbb{E}_w[(\log p(X_i|w)^2] - \mathbb{E}_w[\log p(X_i|w)]^2 \Big\}.
$$

Then $n\mathbb{E}[V_n] \to 2\nu$. The universal laws of Bayesian statistics hold,

$$
\mathbb{E}[G_n] = \mathbb{E}\Big[C_n\Big] + o(\frac{1}{n}), \tag{6.11}
$$

$$
\mathbb{E}[G_n] = \mathbb{E}\Big[W_n\Big] + o(\frac{1}{n}). \tag{6.12}
$$

If a sample is not independent, eq.(6.11) does not hold in general, whereas eq.(6.12) holds in the cases discussed in Section 5.5.

Remark 49. Let W_0 be the set of parameters which minimize the average log density ratio function $K(w)$. The posterior distribution is the summation of local distributions near W_0. However, when $n \to \infty$, the posterior parameters are not to distributed on all neighborhoods of W_0 but restricted on the essential local coordinates. Recall that the essential local coordinates are characterized by the phenomenon that the local real log canonical threshold is minimized. In other words, only the local parameters that have the smallest local real log canonical thresholds are realized by the posterior distribution if n is sufficiently large. If a prior has a hyperparameter, then the essential local coordinates are changed by controlling the hyperparameter (phase transition), which affects the free energy and the generalization loss.

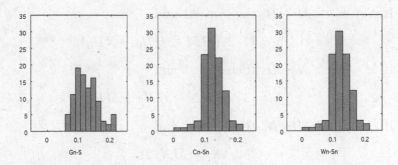

Figure 6.4: For a reduced rank regression model, the generalization, cross validation, and WAIC errors are compared, $G_n - S$, $C_n - S_n$ and $W_n - S_n$. Averages are asymptotically equal to λ/n. Resolution theorem gives the mathematical prediction of the generalization error whose average can be estimated by cross validation and WAIC errors.

It is remarkable that the Bayesian posterior distribution automatically minimizes the free energy and the generalization loss for a given hyperparameter. This is the fundamental mathematical structure of Bayesian statistics.

Example 48. Let $M > 0$, $N > 0$, $H > 0$, and $H_0 \geq 0$ be integers. The reduced rank regression is defined by the statistical model of $y \in \mathbb{R}^N$ for a given $x \in \mathbb{R}^M$

$$p(y|x, A, B) = \frac{1}{(2\pi)^{N/2}} \exp(-\frac{1}{2}\|y - BAx\|^2),$$

where $A = (A_{jk})$ and $B = (B_{k\ell})$ are $N \times H$ and $H \times M$ matrices which have real coefficients respectively. A prior is set

$$\varphi(A, B) \propto \exp(-\frac{1}{2}(\sum_{j,k}(A_{jk})^2 + \sum_{k,\ell}(B_{k\ell})^2)).$$

Assume a true distribution $q(y|x) = p(y|A_0, B_0)$ which satisfies $H_0 = \text{rank}(B_0 A_0)$, $H_0 \leq H$. The complete resolution map was given by [5].
(1) If $N + H_0 < M + H$, $M + H_0 < N + H$, $H + H_0 < M + N$, and $M + N + H + H_0$ is an even integer, then $m = 1$ and

$$\lambda = (1/8)\{2(H + H_0)(M + N) - (M - N)^2 - (H + H_0)^2\}.$$

(2) If $N + H_0 < M + H$, $M + H_0 < N + H$, $H + H_0 < M + N$, and $M + N + H + H_0$ is an odd integer, $m = 2$ and

$$\lambda = (1/8)\{2(H + H_0)(M + N) - (M - N)^2 - (H + H_0)^2 + 1\}.$$

(3) If $N + H_0 > M + H$, then $m = 1$ and

$$\lambda = (1/2)\{HM - HH_0 + NH_0\}.$$

(4) If $M + H_0 > N + H$, then $m = 1$ and

$$\lambda = (1/2)\{HN - HH_0 + MH_0\}.$$

(5) If $H + H_0 > M + N$, then $m = 1$ and

$$\lambda = (MN/2).$$

Note that if a prior satisfies $\varphi(A, B) > 0$, then λ and m are same as above. Figure 6.4 shows experimental results for the generalization, cross validation, and WAIC errors for $M = N = H = 5$, $H_0 = 3$. Then, for $n = 100$, $\lambda/n = 0.12$. The experimental averages and standard deviations were

$$
\begin{aligned}
G_n - S &= 0.126, \quad 0.035, \\
C_n - S_n &= 0.127, \quad 0.034, \\
W_n - S_n &= 0.126, \quad 0.034.
\end{aligned}
$$

It seems that the sample size $n = 100$ is not sufficiently large, however, the theoretical value coincides with the numerical results. From the mathematical point of view, the general theory needs the condition that n is sufficiently large, however, it holds even for smaller n in many concrete statistical models. Hence the real log canonical threshold can be used for checking whether the posterior distribution is accurately approximated by Markov chain Monte Carlo or not.

6.4   Maximum A Posteriori Method

In this section we study the asymptotic property of the maximum *a posteriori* method in nonregular cases. If a prior is defined by the uniform distribution, then it is equivalent to that of the maximum likelihood method. In general, their results are very different from Bayesian estimation, because a single parameter is chosen. In statistical inferences, a single parameter is far from a distribution on a parameter. For example, see Example 60. The average and empirical log loss functions are

$$
\begin{aligned}
L(w) &= -\int q(x) \log p(x|w) dx, \\
L_n(w) &= -\frac{1}{n} \sum_{i=1}^{n} \log p(X_i|w),
\end{aligned}
$$

where $w \in W$. In this section, we study the case that W is a compact set. Let $\hat{w}$ be the parameter that minimizes

$$\mathcal{L}(w) = L(w) - \frac{1}{n} \log \varphi(w),$$

which is called the maximum *a posteriori* (MAP) estimator. If $\log \varphi(w)$ is a constant for all w, then $\hat{w}$ is the maximum likelihood estimator (MLE). The generalization and training losses are respectively defined by

$$\begin{aligned} G_n(MAP) &= L(\hat{w}), \\ T_n(MAP) &= L_n(\hat{w}). \end{aligned}$$

The set of optimal parameters W_0 is defined by

$$W_0 = \{w \in W \ ; \ L(w) \text{ is minimized}\}.$$

We assume that the parameter set W_0 is compact and the convergence in probability holds,

$$\min_{w_0 \in W_0} \|\hat{w} - w_0\| \to 0.$$

Hence we can restrict the parameter set in the union of the neighborhoods of W_0. By using the resolution theorem in each local neighborhood U_j,

$$\begin{aligned} nK_n(g(u)) &= -\sum_{i=1}^{n} \log p(X_i|g(u)) \\ &= nu^{2k} - \sqrt{n}u^k \xi_n(u), \end{aligned}$$

where we can assume $u \in [0,1]^d$ without loss of generality and $k = (k_1, k_2, ..., k_d)$. In this section, the integer r $(1 \le r \le d)$ is defined so that $k_1, k_2, ..., k_r > 0$ and

$$u^{2k} = u_1^{2k_1} u_2^{2k_2} \cdots u_r^{2k_r}.$$

In other words, $\{k_r\}$ does not contain zero. For a given u, let a $(1 \le a \le r)$ be the positive integer which satisfies

$$\frac{u_a^2}{k_a} \le \frac{u_i^2}{k_i} \quad (i = 1, 2, ..., r). \tag{6.13}$$

A map

$$[0,1]^d \ni u \mapsto (t, v) = (t, (v_1, v_2, ..., v_d)) \in \mathbb{R}^1 \times [0,1]^d$$

is defined by

$$t = u^{2k},$$

$$v_i = \begin{cases} \sqrt{u_i^2 - (k_i/k_a)u_a^2} & (1 \le i \le r) \\ u_i & (r < i \le d) \end{cases}.$$

By the definition, $v_a = 0$, hence v is contained in the set,

$$V \equiv \{x = (x_1, x_2, ..., x_d) \in [0,1]^d \; ; \; x_1 x_2 \cdots x_r = 0\}.$$

Moreover, the map

$$[0,1]^d \ni u \mapsto (t,v) \in T \times V$$

is one-to-one.

Example 49. Let us illustrate a case when $u^{2k} = u_1^2 u_2^4$. Then $(k_1, k_2) = (1, 2)$,

$$t = u_1^2 u_2^4,$$

$$v_1 = \sqrt{u_1^2 - (1/2)u_2^2} \quad (\text{if } u_2^2/2 \le u_1^2),$$

$$v_2 = \sqrt{u_2^2 - 2u_1^2} \quad (\text{if otherwise}).$$

Figure 6.5 shows the coodinate (t, v_1) and (t, v_2). The coordinate (t, v_1) is used if $u_2^2/2 \le u_1^2$ whereas (t, v_2) if otherwise.

By using the coordinate (t, v) and Theorem 8, $\mathcal{L}(g(u))$, $L(g(u))$, and $L_n(g(u))$ are represented as the functions of (t, v),

$$\mathcal{L}(t,v) = L_n(w_0) + t - \sqrt{t/n}\, \xi_n(t,v) - \frac{1}{n}\log \varphi(t,v), \qquad (6.14)$$

$$L(t,v) = L(w_0) + t, \qquad (6.15)$$

$$L_n(t,v) = L_n(w_0) + t - \sqrt{t/n}\, \xi_n(t,v). \qquad (6.16)$$

By using these representations, we can derive the average and empirical log losses. Before the theorem, we prepare a lemma.

Lemma 25. *Let $f(u)$ be a C^1-class function of u. There exists $C > 0$ such that*

$$|f(t,v) - f(0,v)| \le C\, t^{1/(2|k|)} |\nabla f| \quad (0 \le t < 1),$$

where $2|k| = 2(k_1 + \cdots + k_r)$ and

$$|\nabla f| = \sup_{u \in [0,1]^d} \max_{1 \le j \le d} \left| \frac{\partial f}{\partial u_j}(u) \right|.$$

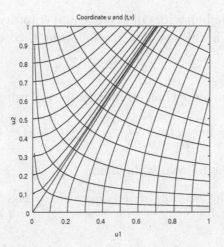

Figure 6.5: Coordinates for MAP and ML. The coordinate for analyzing the maximum a posterior estimator in the case $u^{2k} = u_1^2 u_2^4$. Coordinates are $t = u_1^2 u_2^4$, $v_1 = (u_1^2 - u_2^2/2)^{1/2}$, and $v_2 = (u_2^2 - 2u_1^2)^{1/2}$. The coordinate (t, v_1) is used if $u_2^2/2 \leq u_1^2$ whereas (t, v_2) is used if otherwise.

Proof. For $u = (t, v)$ and $u' = (0, v)$,

$$
\begin{aligned}
|f(t, v) - f(0, v)| &= |f(u) - f(u')| \leq \|u - u'\| \, |\nabla f| \\
&\leq \sqrt{r} \max_j |u_j - u'_j| \, |\nabla f| \\
&\leq C|u^{2k}|^{1/(2|k|)} \, |\nabla f|,
\end{aligned}
$$

where the last inequality is proved as follows. If $j = a$ then $|u_j - u'_j| = u_a$, else if $j \neq a$ then $u_a^2/k_a \leq u_j^2/k_j$. Hence

$$
\begin{aligned}
|u_j - u'_j| &= |u_j - (u_j^2 - (k_j/k_a)u_a^2)^{1/2}| \\
&= \frac{(k_j/k_a)u_a^2}{u_j + (u_j^2 - (k_j/k_a)u_a^2)^{1/2}} \\
&= \frac{(k_j/k_a)u_a^2}{u_j} \leq \sqrt{k_j/k_a} \, u_a.
\end{aligned}
$$

There exists $C' > 0$ such that

$$
(u_a)^{2|k|} \leq C' \, u^{2k},
$$

which completes the lemma. $\qquad\qquad\square$

The asymptotic behaviors of average and empirical log loss functions for the MAP method are derived by the following theorem. Let us define

$$\psi(u) = -\log \varphi(g(u)).$$

Let u^* be the parameter that minimizes the following function,

$$u^* = \arg \min_{K(g(u))=0} \left\{ -\frac{1}{4} \min\{0, \xi_n(u)\}^2 + \psi(u) \right\}. \qquad (6.17)$$

Then we obtain the following theorem.

Theorem 18. *The average and empirical log loss functions of the MAP method satisfy*

$$L(\hat{w}) = L(w_0) + \frac{1}{4n} \max\{0, \xi_n(u^*)\}^2 + o_p(\frac{1}{n}),$$

$$L_n(\hat{w}) = L_n(w_0) - \frac{1}{4n} \max\{0, \xi_n(u^*)\}^2 + o_p(\frac{1}{n}).$$

Proof. By applying Lemma 25 to $\xi_n(t, v)$, if $t \to 0$, then

$$\xi_n(t, v) - \xi_n(0, v) = o_p(1).$$

The eq.(6.14) is equal to

$$\mathcal{L}(t, v) = \left(\sqrt{t} - \frac{\xi_n(t, v)}{2\sqrt{n}} \right)^2 - \frac{\xi_n(t, v)^2}{4n} + \frac{\psi(t, v)}{n} + L_n(w_0). \qquad (6.18)$$

Let $(\hat{t}, \hat{v})$ be the set of parameters that minimizes $\mathcal{L}(t, v)$. Firstly we study the case $\xi_n(\hat{t}, \hat{v}) \leq 0$. Then there exists $t^* = O_p(1)$ such that

$$\sqrt{\hat{t}} = \frac{t^*}{\sqrt{n}},$$

because, if this equation holds, the order of $\mathcal{L}(\hat{t}, \hat{v}) - L_n(w_0)$ is not larger than $(1/n)$, if otherwise, it is larger than $1/n$. Then

$$\mathcal{L}(\hat{t}, \hat{v}) = \frac{1}{n} \left\{ (t^* - \frac{\xi_n(0, \hat{v})}{2})^2 - \frac{\xi_n(0, \hat{v})^2}{4} + \psi(0, \hat{v}) \right\}$$
$$+ L_n(w_0) + o_p(\frac{1}{n}).$$

Since $\xi_n(\hat{t}, \hat{v}) \leq 0$ and $\hat{t} = O_p(1/n)$, $\xi_n(0, \hat{v}) < o_p(1)$. Therefore $\mathcal{L}(\hat{t}, \hat{v})$ is minimized by $t^* = o_p(1)$ and

$$\hat{v} = \mathrm{argmin}_v \psi(0, v) + o_p(1).$$

Thus $\hat{t} = o_p(1/n)$ and eqs.(6.15) and (6.16),

$$
\begin{aligned}
L(\hat{w}) &= L(w_0) + o_p(1/n), \\
L_n(\hat{w}) &= L_n(w_0) + o_p(1/n).
\end{aligned}
$$

Secondly, we study the case $\xi_n(\hat{t}, \hat{v}) > 0$. There exists $t^* = O_p(1)$ such that

$$
\sqrt{\hat{t}} = \frac{1}{2\sqrt{n}}(\xi_n(\hat{t}, \hat{v}) + t^*) = \frac{1}{2\sqrt{n}}(\xi_n(0, \hat{v}) + t^*) + o_p(1/\sqrt{n}),
$$

because, if otherwise, the order of $\mathcal{L}(\hat{t}, \hat{v})$ is larger than $(1/n)$ by eq.(6.18). Then also by eq.(6.18) and $\hat{t} = o_p(1)$,

$$
\mathcal{L}(\hat{t}, \hat{v}) = \frac{(t^*)^2}{4n} - \frac{\xi_n(0, \hat{v})^2}{4n} + \frac{\psi(0, \hat{v})}{n} + L_n(w_0) + o_p(1/n). \tag{6.19}
$$

Hence $t^* = o_p(1)$ and

$$
\hat{v} = \operatorname{argmin}_v(-\xi_n(0, v)^2 + 4\psi(0, v)) + o_p(1).
$$

Therefore,

$$
\begin{aligned}
L(\hat{w}) &= L(w_0) + \frac{1}{4n}\xi_n(0, \hat{v})^2 + o_p(1/n), \\
L_n(\hat{w}) &= L_n(w_0) - \frac{1}{4n}\xi_n(0, \hat{v})^2 + o_p(1/n).
\end{aligned}
$$

By integrating the above two cases, the theorem is obtained. $\square$

Theorem 19. *There exists $\mu > 0$ such that*

$$
\begin{aligned}
\mathbb{E}[L(\hat{w})] &= L(w_0) + \frac{\mu}{n} + o(\frac{1}{n}), \\
\mathbb{E}[L_n(\hat{w})] &= L_n(w_0) - \frac{\mu}{n} + o(\frac{1}{n}).
\end{aligned}
$$

Proof. The parameter u^* is defined by minimization,

$$
u^* = \operatorname{argmin}_u\Big\{-\frac{1}{4}\min_{K(g(u))=0}\{0, \xi_n(u)\}^2 + \psi(u)\Big\}. \tag{6.20}
$$

By using the convergence of the empirical process $\xi_n(u) \to \xi(u)$, and its average in Sections 10.4 and 10.5, this theorem is obtained by

$$
\mu = \lim_{n \to \infty} \mathbb{E}[\max\{0, \xi_n(u^*)\}^2]/4.
$$

$\square$

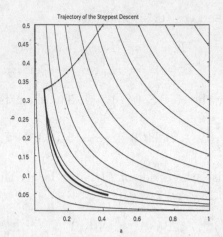

Figure 6.6: Trajectory of steepest descent. The contour of the square error of a simple neural network and trajectory by the steepsest descent are shown. The parameter has two representations (a, b) and (t, v), where $t = a^2 b^2$. Optimization about t makes the generalization error smaller, whereas that about v larger.

Example 50. (Over-training) Let us study a simple neural network of x, y, a, b ℝ,

$$p(y|x, a, b) = \frac{1}{\sqrt{2\pi\sigma^2}} \exp\left(-\frac{1}{2\sigma^2}(y - a\tanh(bx))^2\right),$$

where $\sigma = 0.1$. The random variable X is subject to the uniform distribution on $[-2, 2]$ and the true conditional density $q(y|x) = p(y|x, 0, 0)$. The contour of the square error

$$E(a, b) = \frac{1}{n} \sum_{i=1}^{n} (Y_i - a\tanh(bX_i))^2$$

and the trajectory by the steepest descent are shown in Figure 6.6. In this case $t = a^2 b^2$. The parameter can be represented by (a, b) and (t, v). In the steepest descent, the parameter t is rapidly optimized, whereas v is searched very slowly. Optimization about v gives the over-training, that is to say, the generalization error is made smaller by optimization about t, but larger about v. If Bayesian estimation is employed, then the posterior distribution is spread over $\{v\}$ and t is optimized, which makes the generalization smaller. This is the main difference between MAP or ML and Bayesian estimation.

Remark 50. (1) The constant μ depends on the prior $\varphi(w)$. In the maximum likelihood method, it is given by the average of the maximum value of the Gaussian process. If W is compact, it is finite but not small in general. If W is not compact, the maximum value of the Gaussian process is not finite in general, hence the asymptotic properties of the above theorem do not hold. The maximum likelihood method is not appropriate if the likelihood function cannot be approximated by normal distribution.

(2) In the above theorems, we study the maximum *a posteriori* (MAP) method on the parameter $w \in W$. In order to study MAP method on $u \in \cup U_j$, then $\varphi(g(u))$ should be replaced by $\varphi(g(u))|g'(u)|$. Note that MAP depends on the transform of the parameter. In other words, MAP is not invariant about parameter representation, whereas ML and Bayes are invariant.

(3) In general, the average parameter $\mathbb{E}_w[w]$ does not converge to W_0 in normal mixture and neural networks. Hence

$$L(\mathbb{E}_w[w]) = nL(w_0) + nC + O_p(1).$$

In other words, the posterior mean estimator is not appropriate if the posterior distribution cannot be approximated by a normal distribution.

6.5 Problems

1. For a given analytic funciton $K(w) \geq 0$ and a prior $\varphi(w)$, the partition function, the state density function, and the zeta function are defined by

$$
\begin{aligned}
Z(n) &= \int \exp(-nK(w))\varphi(w)dw \quad (n > 0), \\
v(t) &= \int \delta(t - K(w))\varphi(w)dw \quad (t > 0), \\
\zeta(z) &= \int K(w)^z \varphi(w)dw \quad (z \in \mathbb{C}).
\end{aligned}
$$

Show that the following are equivalent.

(1) If $n \to \infty$, then $\log Z(n) + \lambda \log n \to 0$.

(2) If $t \to +0$, then $v(t)/t^{\lambda-1} \to c > 0$ for some c.

(3) $\zeta(z)$ is holomorphic in the region $\text{Re}(z) > -\lambda$ and has a pole at $z = -\lambda$ with the order 1.

2. Assume that two sets of analytic functions and priors $(K_1(w), \varphi_1(w))$ and $(K_2(w), \varphi_2(w))$ have the real log canonical thresholds λ_1 and λ_2, re-

spectively. Also assume that $K_1(w) \geq cK_2(w)$ for some $c_1 > 0$ and that $\varphi_1(w) \leq c_2\varphi_2(w)$ for some $c_2 > 0$. Then prove that $\lambda_1 \geq \lambda_2$.

3. Assume that two sets of analytic functions and priors $(K_1(w_1), \varphi_1(w_1))$ and $(K_2(w_2), \varphi_2(w_2))$ have the real log canonical thresholds λ_1 and λ_2, respectively, where w_1 and w_2 are different variables. Then prove the following.
(1) The real log canonical threshold of $(K_1(w_1) + K_2(w_2), \varphi_1(w_1)\varphi_2(w_2))$ is $\lambda_1 + \lambda_2$.
(2) The real log canonical threshold of $(K_1(w_1)K_2(w_2), \varphi_1(w_1)\varphi_2(w_2))$ is $\min\{\lambda_1, \lambda_2\}$.

4. Assume that two sets of functions $\{f_j(w); j = 1, 2, ..., J\}$ and $\{g_k(w); k = 1, 2, ..., K\}$ satisfy

$$\left\{ \sum_{j=1}^{J} a_j(w)f_j(w); a_j(w) \in \mathcal{R} \right\} = \left\{ \sum_{k=1}^{K} b_k(w)g_k(w); b_j(w) \in \mathcal{R} \right\}, \quad (6.21)$$

where $\mathcal{R} = \mathbb{R}[w_1, w_2, ..., w_d]$ is the polynomial ring generated by 1, w_1, w_2, ...,w_d with the real coefficients. Then prove that the following two functions have the same real log canonical thresold if the same prior is employed.

$$\sum_{j=1}^{J} (f_j(w))^2, \quad \sum_{k=1}^{K} (g_k(w))^2.$$

Note that if eq.(6.21) holds, it is said that the ideal generated by $\{f_j(w)\}$ is equal to that of $\{g_j(w)\}$.

5. Let us define a function $\overline{F}_n$ by

$$\overline{F}_n = -\log \int \exp(-nL(w))\varphi(w), \quad (6.22)$$

where $L(w)$ is the average log loss function. Then prove that

$$\overline{F}_n = nL(w_0) + \lambda \log n - (m - 1) \log \log n + O(1).$$

Hence the difference between F_n and $\overline{F}_n$ is a constant order random variable.

6. For a general random variable X, $\mathbb{E}[\exp(X)] \geq \exp \mathbb{E}[X]$ holds, which is called Jensen's inequality. Prove that for a general random variables X and Y and a general function $f(X, Y)$,

$$\mathbb{E}_X[-\log \mathbb{E}_Y[\exp(f(X, Y))]] \leq -\log \mathbb{E}_Y[\exp(\mathbb{E}_X[f(X, Y)])]$$

by using Jensen's inequality. By using this inequality, prove that

$$F_n \leq \overline{F}_n,$$

where $\overline{F}_n$ is defined by eq.(6.22).

Chapter 7

Markov Chain Monte Carlo

The Markov chain Monte Carlo method (MCMC) enables us to numerically approximate the posterior average for an arbitrary statistical model and prior. If a posterior distribution is spread on some local parameter region, then MCMC approximation is accurate, otherwise it is still not so easy. In many important statistical models such as a normal mixtture or an artificial neural network, the Bayesian inference attains much more precise estimation, hence it becomes more important to construct the MCMC algorithm which works even in singular posterior distributions. In this chaper, we introduce the basic foundations of MCMC process.

(1) The Metropolis method is explained. The Hamiltonian Monte Carlo and the parallel tempering are its advanced versions.

(2) The Gibbs sampler is introduced. Nonparametric Bayesian sampler is its advanced version.

(3) Numerical approximation methods of the generalization loss and the free energy using MCMC method are explained.

In order to check how accurate MCMC approximates the posterior distribution in singular cases, the real log canonical threshold would be a good index for a given set of a true distribution, a statistical model, and a prior.

7.1 Metropolis Method

Let $p(x|w)$ and $\varphi(w)$ be a statistical model and a prior, where $x \in \mathbb{R}^N, w \in W \subset \mathbb{R}^d$. The Hamiltonian function $H(w)$ is defined by

$$H(w) = -\sum_{i=1}^{n} \log p(X_i|w) - \log \varphi(w).$$

Then the posterior distribution is represented by

$$p(w) = \frac{1}{Z_n}\varphi(w)\prod_{i=1}^{n}p(X_i|w)$$

$$= \frac{1}{Z_n}\exp(-H(w)),$$

where Z_n is a partition function or the marginal likelihood,

$$Z_n = \int \exp(-H(w))dw.$$

The probability density function $p(w)$, which is equal to the posterior distribution in Bayesian statistics, is called the equilibrium state of the Hamiltonian $H(w)$. Our purpose in this chapeter is to generate $\{w_k\}_{k=1}^{K}$ such that, for an arbitrary function $f(w)$,

$$\int f(w)p(w)dw \approx \frac{1}{K}\sum_{k=1}^{K}f(w_k), \tag{7.1}$$

when $K \to \infty$. In most cases in statistical applications, n is large, hence the parameter set

$$\{w \in W; p(w) > \epsilon\}$$

for some $\epsilon > 0$ is very a narrow subset of W, resulting that Riemann sum of the integral on the parameter space does not give effective approximation.

Remark 51. (1) (Curse of dimensionality) If $d = 1$, then the integral can be approximated by Reimann sum,

$$\int_0^1 f(w)p(w)dw \approx \frac{1}{K}\sum_{k=1}^{K}f(k/K)p(k/K), \tag{7.2}$$

which is more accurate than MCMC. However, if $d = 2, 3, 4, ...$, then the number K^d necessary for approximation becomes too large to be calculated numerically. This difficulty is called "curse of dimensionality".

(2) (Importance sampling) If a function $H_0(w)$ exists such that $\{w_k\}$ can be easily generated from $p_0(w) \propto \exp(-H_0(w))$, then

$$\int f(w)p(w)dw \approx \frac{\displaystyle\sum_{k=1}^{K}f(w_k)\exp(-H(w_k) + H_0(w_k))}{\displaystyle\sum_{k=1}^{K}\exp(-H(w_k) + H_0(w_k))}. \tag{7.3}$$

This method is called the importance sampling, which works well if $H(w) \approx H_0(w)$.

(3) In almost all cases, $H(w)$ is given explicitly, however, Z_n not. It is more difficult to calculate Z_n than estimating the average.

In the Markov chain Monte Carlo (MCMC) method, a sequence $\{w_1, w_2, w_3, ...\}$ is generated by a conditional probability $p(w_{k+1}|w_k)$ iteratively. It is known that (1) and (2) below are sufficient conditions for eq.(7.1) to hold.

(1) (**Detailed Balance Condition**). For arbitrary parameters w_a, $w_b \in W$,

$$p(w_b|w_a)p(w_a) = p(w_a|w_b)p(w_b).$$

(2) (**Irreducible Condition**). For an arbitrary $w \in W$, the probability that a parameter of $\{w_k\}$ is contained in the neighborhood of w is not equal to zero.

Note that the detailed balance condition is not necessary for eq.(7.1), however, there are several MCMC algorithms which satisfy the detailed balance condition. Firstly, we study Metropolis method.

7.1.1   Basic Metropolis Method

Let $r(w_1|w_2)$ be a conditional probability density which satisfies

$$\forall(w_1, w_2), \quad r(w_1|w_2) = r(w_2|w_1). \tag{7.4}$$

In Metropolis method, the set of parameters $\{w(t) \in \mathbb{R}^d; t = 1, 2, 3, ...\}$ is generated as follows.

Metropolis Method.

(1) Initialize $w(1)$ and $t = 1$.

(2) A candidate w' is generated by $r(w'|w(t))$.

(3) By using $\Delta H \equiv H(w')-H(w(t))$, the probability $P = \min\{1, \exp(-\Delta H)\}$ is determined. Then set $w(t+1) = w'$ with probability P, or $w(t+1) = w(t)$ with $1 - P$.

(4) $t := t + 1$, and return to (2).

We can prove that this procedure satisfies the detailed balance condition.

Theorem 20. *Metropolis method satisfies the detailed balance condition.*

Proof. Let $p(w(t+1)|w(t))$ be the conditional probability which is used in one step of the Metropolis method. To show the detailed balance condition, it is sufficient to prove

$$p(w_a|w_b)\exp(-H(w_b)) = p(w_b|w_a)\exp(-H(w_a)) \qquad (7.5)$$

for an arbitrary set (w_a, w_b). For a given $w(t)$, the simultaneous probability that w' is generated from $w(t)$ and that $w(t+1) = w'$ is

$$r(w'|w(t))\min\{1, \exp(-H(w') + H(w(t)))\}.$$

Therefore, for a given $w(t)$, the probability that the new candidate place is chosen is given by marginalization about w',

$$Q(w(t)) = \int r(w'|w(t))\min\{1, \exp(-H(w') + H(w(t)))\}dw'.$$

Hence the probability that $w(t+1) = w(t)$ is $1 - Q(w(t))$, resulting that

$$
\begin{aligned}
p(w_a|w_b) &= r(w_a|w_b)\min\{1, \exp(-H(w_a) + H(w_b))\} \\
&\quad + \delta(w_a - w_b)(1 - Q(w_b)).
\end{aligned}
$$

By using this relation, $r(w_a|w_b) = r(w_b|w_a)$, and the property of the delta function, it follows that

$$
\begin{aligned}
p(w_a|w_b)\exp(-H(w_b)) &= r(w_a|w_b)\min\{\exp(-H(w_b)), \exp(-H(w_a))\} \\
&\quad + \delta(w_a - w_b)(1 - Q(w_b))\exp(-H(w_b)) \\
&= r(w_b|w_a)\min\{\exp(-H(w_b)), \exp(-H(w_a))\} \\
&\quad + \delta(w_b - w_a)(1 - Q(w_a))\exp(-H(w_a)) \\
&= p(w_b|w_a)\exp(-H(w_a)),
\end{aligned}
$$

which completes the theorem.   □

Remark 52. (Metropolis-Hasting method) Metropolis method can be generalized for the case $r(w_a|w_b) \neq r(w_b|w_a)$. For such a case, the probability P is replaced by

$$P = \min\left\{1, \frac{r(w(t)|w')\exp(-H(w'))}{r(w'|w(t))\exp(-H(w(t)))}\right\}.$$

Then the detailed balance condition is satisfied.

Remark 53. Theoretically speaking, Metropolis method gives the set of parameter which ensures eq.(7.1), if $K \to \infty$. However, in practical applications, there are several issues.

(1) The parameters in the period which is affected by the initial point should be removed from the obtained parameter set. Such a period is called 'burn-in'.

(2) The obtained parameters are not independent if MCMC is used. For the effective approximation of the posterior distribution, the dependency between parameters had better be reduced. Hence $\{w(m*t); t = 1, 2, ...\}$ for some m is chosen. If m is large, then dependency of the obtained parameters is made small, but it needs a computational cost. In this book, m is called a 'sampling interval'.

(3) If a probability distribution $p(w)$ has several distant peaks, then the probability from a peak to another peak becomes very small, hence the irreducibility of MCMC often fails. This is called the problem of a 'potential barrier'.

(4) Let w_0 be the parameter that minimizes $H(w)$. If the set $\{w \in W; H(w) - H(w_0) < \epsilon\}$ is connected but not contained in some local region, then MCMC process sometimes fails because the probability from a place to a distant place is very small. This is called the problem of 'entropy barrier'.

(5) The probability that the candidate parameter w' is chosen is called the acceptance probability. If the variance of $r(w_1|w_2)$ is small, then the acceptance probability becomes high, but the candidate parameter is chosen in the narrow local region. If it is large, then the acceptance probability becomes small, but the candidate parameter is chosen from the wide range. In the Metropolis method, optimization of the acceptance probability by controlling $r(w_1|w_2)$, is one of the most important processes for constructing MCMC.

(6) Several criteria which judge whether the parameters could be understood as taken from the equilibrium state or not are proposed [29, 30, 33].

7.1.2 Hamiltonian Monte Carlo

In the original Metropolis method, in order to ensure the acceptance probability is not small, the dependency of $w(t)$ and $w(t+1)$ becomes large. The following method was devised to improve this property.

Let $w \in \mathbb{R}^d$. A new variable $v \in \mathbb{R}^d$ is introduced and the total Hamiltonian of (w, v) is defined by

$$\mathcal{H}(w, v) = \frac{1}{2}\|v\|^2 + H(w).$$

If $\{(w_k, v_k)\}$ which is subject to the equilibrium state $\exp(-\mathcal{H}(u, v))$ of the total Hamiltonian, then $\{w_k\}$ is subject to the equilibrium state $\exp(-H(u))$ of the Hamiltonian $H(w)$. Thus we make $\{(w_k, v_k)\}$ subject to the equilibrium state of the total Hamiltonian.

Hamiltonian Monte Carlo.

(1) Initialize $w(1)$ and $t = 1$.

(2) The elements of $v \in \mathbb{R}^d$ are independently generated by the standard normal distribution.

(3) The following differential equation with respect to the time parameter τ is solved with the initial condition that $(w(t), v)$ at $\tau = 0$. Here τ is a variable which has no relation to MCMC time t.

$$\frac{dw}{d\tau} = v,$$
$$\frac{dv}{d\tau} = -\nabla H(w).$$

This is known as the Hamilton equation which describes the dynamics of the canonical coordinate (u, v). Then (w', v') $(w' = w(\tau), u' = u(\tau))$ is obtained for a given time τ. It is permissible that the numerical solution of the differential equation contains errors. However, it should satisfy the invariance condition of the time reverse and the volume conservation condition of the phase space. It is known that leap frog method in Remark 54 satisfies both conditions.

(4) By defining $\Delta\mathcal{H} = \mathcal{H}(w', v') - \mathcal{H}(w(t), v)$, then $w(t + 1) = w'$ with $P = \max\{1, \exp(-\Delta\mathcal{H})\}$, or $w(t + 1) = w(t)$ with probability $1 - P$.

(5) $t = t + 1$. Return to (2).

This method satisfies the detailed balance condition for $\exp(-\mathcal{H}(w, p))$. Note that the rigorous solution of the differential equation satisfies $d\mathcal{H}/dt = 0$, hence it is expected that the numerical solution gives $\Delta\mathcal{H} \approx 0$, thus the acceptance probability can be made higher and w' can be generated at the distant place from $w(t)$.

Remark 54. (Leap frog method) The differential equation

$$\frac{dw}{dt} = v, \quad \frac{dv}{dt} = f(w)$$

is numerically solved by the iteration,

$$v(n+1/2) = v(n) + \frac{\epsilon}{2} f(w(n)),$$
$$w(n+1) = w(n) + \epsilon\, v(n+1/2),$$
$$v(n+1) = v(n+1/2) + \frac{\epsilon}{2} f(w(n+1)),$$

where $\epsilon > 0$ is a small constant.

If ϵ is made very small, then the differential equation is solved with high accuracy but the computational costs also become high. In Hamiltonian Monte Carlo, the controlling the balanace between ϵ and the acceptance probability is necessary. Recently, an improved algorithm was proposed which determines them automatically by using non-U-turn Hamiltonian Monte Carlo [38].

Example 51. For a probability density

$$p(x,y) \propto \exp(-Nx^2y^2 - Mx^2 - My^2),$$

where $N = 50, M = 0.005$, random variables $\{(X_i, Y_i)\}$ $(i = 1, 2, ..., 300)$ are generated by Metropolis method, Gibbs sampling, and Hamiltonian Monte Carlo. (See Figure 7.1.) For the Gibbs sampler, see the following subsection. Note that, if $M = 0$, then $\int \exp(-Nx^2y^2)dxdy = \infty$, hence $p(x,y)$ is not a probability density. The origin is a singularity of X^2y^2. In a normal mixture or an artificial neural network, such a singular posterior distribution on higher parameter space is necessary. By an experiment in which 10000 random variables are generated, the empirical averages are compared,

$$\mathbb{E}_{MET}[X] = 0.076,$$
$$\mathbb{E}_{MET}[Y] = 0.4528,$$
$$\mathbb{E}_{GIB}[X] = 0.0289,$$
$$\mathbb{E}_{GIB}[Y] = -0.017,$$
$$\mathbb{E}_{HAM}[X] = 0.091,$$
$$\mathbb{E}_{HAM}[Y] = 0.089,$$

where $\mathbb{E}_{MET}$, $\mathbb{E}_{GIB}$, and $\mathbb{E}_{HAM}$ mean the empirical averages oy the Metropolis method, Gibbs sampler, and Hamiltonian Monte Carlo. The true averages of X and Y are equal to zero. In the Metropolis and Hamiltonian methods, one sampling process consists of 100 trials and 100 dynamical calculations, respectively. In this case, the Hamiltonian has the form for which Gibbs sampler can be employed. However, in general it is not applied. In the cases when Gibbs sampling cannot be employed, Hamiltonian Monte Carlo gives the more accurate MCMC expectations.

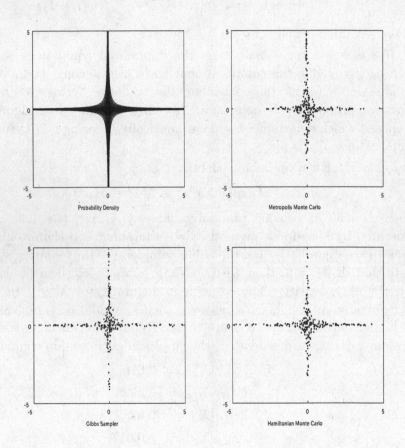

Figure 7.1: Comparison of a probability distribution, Metropolis method, Gibbs Sampler, and Hamiltonian method for $p(x,y) \propto \exp(-Nx^2y^2 - Mx^2 - My^2)$. In many statistical models, the posterior distribution contains singularities, hence MCMC processes for such cases are very important.

7.1.3 Parallel Tempering

If the posterior distribution does not concentrate in some local region, the parallel tempering or replica Monte Carlo is sometimes employed. Let

$$nL_n(w) = -\sum_{i=1}^{n} \log p(X_i|w).$$

Note that this function does not contain the prior information, in other words,

$$H(w) = nL_n(w) - \log \varphi(w).$$

The equilibrium state of the inverse temperature $\beta > 0$ is defined by

$$p(w|\beta) = \frac{1}{Z(\beta)} \exp(-n\beta L_n(w))\varphi(w).$$

Then the posterior distribution is equal to $p(w|1)$. Let the sequence of inverse termperatures be

$$0 - \beta_1 < \beta_2 < \cdots < \beta_J = 1.$$

The target probability distribution of the parallel tempering is

$$\prod_{j=1}^{J} p(w_j|\beta_j), \tag{7.6}$$

which is a probability density function of $(w_1, w_2, ..., w_J)$. If parameters are taken from this distribution, then the set $\{w_J(t)\}$ can be used for posterior distribution, because $p(w_J|\beta_J) = p(w|1)$. The parallel tempering consists of two MCMC processes.

Parallel Tempering.

(1) One is the independent MCMC process for each $p(w_j|\beta_j)$.

$$w_{11} \rightarrow w_{12} \rightarrow w_{13} \rightarrow \cdots$$
$$w_{21} \rightarrow w_{22} \rightarrow w_{23} \rightarrow \cdots$$
$$\cdots$$
$$w_{J1} \rightarrow w_{J2} \rightarrow w_{J3} \rightarrow \cdots$$

In this process, arbitrary MCMC method can be used.

(2) The other is the exchange process between w_j and w_{j+1} with some interval in each MCMC process. The probability of the exchange is given by

$$\min\{1, \exp\{(\beta_{j+1} - \beta_j)(nL_n(w_{j+1}) - nL_n(w_j))\}, \qquad (7.7)$$

which satisfies the detailed balance condition for the probability density of eq.(7.6). Note that the prior does not affect the exchange probability.

Even if $p(w_j|\beta_J)$ have many peaks, $p(w_j|\beta_j)$ for small β_j does not, hence the equilibrium state can be more easily realized by exchanging parameters.

Theorem 21. *Parallel tempering using the exchange probability of eq.(7.7) satisfies the detailed balance condition.*

Proof. Let us use (u, v) for (w_i, w_{j+1}) and (α, β) for (β_j, β_{j+1}). The target probability distribution is

$$P(u, v) \ \propto \ \exp(-\alpha n L_n(u))\varphi(u) \exp(-\beta n L_n(v)\varphi(v))$$
$$= \ \exp(-f(x, y)),$$

where

$$f(u, v) = \alpha n L_n(u) + \beta n L_n(v) - \log\varphi(u) - \log\varphi(v).$$

By the exchange $(u, v) \rightarrow (v, u)$,

$$
\begin{aligned}
\Delta f \ &= \ f(v, u) - f(u, v) \\
&= \ \alpha n L_n(v) + \beta n L_n(u) - \alpha n L_n(u) - \beta n L_n(v) \\
&= \ (\beta - \alpha)(n L_n(u) - n L_n(v)).
\end{aligned}
$$

Therefore the exchange process whose probability is defined by

$$\min\{1, \exp(-\Delta f)\}$$

satisfies the detailed balance condition. $\qquad\qquad\square$

Remark 55. Assume that the posterior distribution has the real log canonical threshold λ. The exchange probability between $\beta_1, \beta_2, (\beta_2 > \beta_1)$ is asymptotically (for $n \rightarrow \infty$) given by the following formula, [52].

$$P(\beta_1, \beta_2) = 1 - \frac{1}{\sqrt{\pi}} \frac{\beta_2 - \beta_1}{\beta_1} \frac{\Gamma(\lambda + 1/2)}{\Gamma(\lambda)}.$$

If the sequence $\{\beta_j\}$ is set as a geometric progression, the exchange probability becomes a constant for a sufficiently large n.

7.2 Gibbs Sampler

Metropolis method can be applied to any Hamiltonian function, however, it is not easy to generate parameters globally. Hamiltonian Monte Carlo improved that difficulty. Although the Gibbs sampler can be used in the special posterior distributions, if it can be employed, it is rather easy to generate parameters globally.

In the Gibbs sampler, a parameter $w \in \mathbb{R}^d$ is divided as $w = (w_1, w_2)$. Let the posterior distribution be $p(w_1, w_2)$. Then two conditional probability distributions $p(w_1|w_2)$ and $p(w_2|w_1)$ are defined from $p(w_1, w_2)$. The set of parameters $\{w(t) = (w_1(t), w_2(t)) \in \mathbb{R}^d; t = 1, 2, 3, ...\}$ is generated by the following procedure.

Gibbs Sampler.

(1) Initialize $w(1) = (w_1(1), w_2(t))$. $t = 1$.
(2) One of (A) or (B) is chosen with probability $1/2$.
    (A) w_2' is generated by $p(w_2'|w_1(t))$, then w_1' is generated by $p(w_1'|w_2')$.
    (B) w_1' is generated by $p(w_1'|w_2(t))$, then w_2' is generated by $p(w_2'|w_1')$.
(3) Set $w_{t+1} = (w_1', w_2')$ and $t := t + 1$. Return to (2).

Theorem 22. *Gibbs sampler satisfies the detailed balance condition.*

Proof. The probability density of (w_1', w_2') for a given set (w_1, w_2) is given by

$$p(w_1', w_2'|w_1, w_2) = \frac{1}{2}\{p(w_2'|w_1')p(w_1'|w_2) + p(w_1'|w_2')p(w_2'|w_1)\}.$$

Let us prove the detailed balance condition,

$$p(w_1', w_2'|w_1, w_2)p(w_1, w_2) = p(w_1, w_2|w_1', w_2')p(w_1', w_2').$$

By using the definition of the conditional probability,

$$
\begin{aligned}
p(w_2'|w_1')p(w_1'|w_2)p(w_1, w_2) &= \frac{p(w_1', w_2')}{p(w_1')}\frac{p(w_1', w_2)}{p(w_2)}p(w_1, w_2) \\
&= p(w_1', w_2')\frac{p(w_1', w_2)}{p(w_1')}\frac{p(w_1, w_2)}{p(w_2)} \\
&= p(w_1', w_2')p(w_2|w_1')p(w_1|w_2)
\end{aligned}
$$

By the same method,

$$p(w_1'|w_2')p(w_2'|w_1)p(w_1, w_2) = p(w_1', w_2')p(w_1|w_2')p(w_2|w_1).$$

By the sum of these equations, this theorem is completed. $\square$

Remark 56. (1) In the above definition, the order of sampling w_1 and w_2 is chosen by the same probabilities $1/2$. If the order is fixed, then the detailed balance condition is not satisfied, however, the same equilibrium state can be obtained.

(2) If one of two procedures $p(w_1'|w_2)$ and $p(w_1'|w_2)$ is chosen with the same probability, then the conditional probability is given by

$$p(w_1', w_2'|w_1, w_2) = \frac{1}{2}\{p(w_1'|w_2)\delta(w_2' - w_2) + p(w_2'|w_1)\delta(w_1' - w_1)\},$$

which also satisfies the detailed balance condition.

7.2.1 Gibbs Sampler for Normal Mixture

Gibbs sampler is often employed in the mixture models. Let us derive an algorithm by Gibbs sampler for a normal mixture. In this subsection, a normal distribution of $x \in \mathbb{R}^M$ for a given $b \in \mathbb{R}^M$ is denoted by

$$N(x|b) = \frac{1}{(2\pi)^{M/2}} \exp(-\frac{\|x - b\|^2}{2}).$$

Then a normal mixture is defined by

$$p(x|a, b) = \sum_{k=1}^{K} a_k N(x|b_k),$$

where $a = (a_1, a_2, ..., a_K)$ and $b = (b_1, b_2, ..., b_K)$ are parameters of a normal mixture, which satisfies $\sum a_j = 1$ and $a_j \geq 0$, and $b_k \in \mathbb{R}^d$. For the prior, we adopt

$$\varphi(a) = \frac{1}{z_1} \prod_{k=1}^{K} (a_k)^{\alpha_k - 1},$$

$$\varphi(b) = \frac{1}{z_2} \prod_{k=1}^{K} \exp(-\frac{1}{2\sigma^2}\|b_k\|^2),$$

where $\varphi(a)$ and $\varphi(b)$ are the Dirichlet distribution with index $\{\alpha_k\}$ and the normal distribution respectively. Here $\{\alpha_k\}$ and $\sigma^2 > 0$ are hyperparameters and $z_1, z_2 > 0$ are constants. Let $y = (y^{(1)}, y^{(2)}, ..., y^{(K)})$ be a competitive variable, in other words, y takes value on the following set,

$$\mathcal{C}_K = \{(1, 0, ..., 0), \quad (0, 1, 0, ..., 0), \quad ...(0, 0, 0, ..., 1)\}.$$

Then a statistical model for the simultaneous probability density of (x, y) is defined by

$$p(x, y|a, b) = \prod_{k=1}^{K} \left\{ a_k N(x|b_k) \right\}^{y^{(k)}}.$$

It follows that

$$p(x|a, b) = \sum_{y \in \mathcal{C}_K} p(x, y|a, b).$$

Therefore a normal mixture $p(x|a, b)$ can be understood as a statistical model $p(x, y|a, b)$ which has a latent or hidden variable $y \in \mathcal{C}_K$.

Let $x^n = \{x_1, x_2, ..., x_n\}$ be an independent sample and $y_i \in \mathcal{C}_K$ be the competitive variable which corresponds to a sample point x_i. We use a notation $y^n = \{y_1, y_2, ..., y_n\}$. The kth element of y_i is denoted by $y_i^{(k)}$. Then Bayesian simultaneous probability is

$$p(a, b, x^n, y^n) = \varphi(a)\varphi(b) \prod_{i=1}^{n} p(x_i, y_i|a, b) \tag{7.8}$$

$$= \frac{1}{z_1 z_2} \prod_{k=1}^{K} \left[a_k^{\alpha_k - 1} \exp(-\frac{\|b_k\|^2}{2\sigma^2}) \prod_{i=1}^{n} \left\{ a_k N(x_i|b_k) \right\}^{y_i^{(k)}} \right] \tag{7.9}$$

$$= \frac{1}{Z} \left[\prod_{k=1}^{K} a_k^{\alpha_k - 1 + n_k} \right] \left[\prod_{k=1}^{K} \exp(-H_k(b_k)) \right], \tag{7.10}$$

where Z is a normalizing constant and

$$n_k = \sum_{i=1}^{n} y_i^{(k)},$$

$$H_k(b_k) = \frac{\|b_k\|^2}{2\sigma^2} + \sum_{i=1}^{n} \frac{y_i^{(k)}}{2} \|x_i - b_k\|^2.$$

In Bayesian estimation, we need the posterior parameters $\{(a, b)\}$ which are subject to $p(a, b|x^n)$. If $\{(a, b, y^n)\}$ are subject to $p(a, b, y^n|x^n)$, then $\{(a, b)\}$ of them can be used for numerical approximation of the posterior distribution.

Therefore it is sufficient to make a Gibbs sampler for (a, b) and y^n, in which the conditional probabilities we need are

$$p(y^n|a, b, x^n), \quad p(a, b|x^n, y^n).$$

In fact, by using these conditional probabilities, we can make a Gibbs sampler for $(a, b) \mapsto y^n$ and $y^n \mapsto (a, b)$. By eq.(7.9), under the probability distribution $p(y^n | a, b, x^n)$, $y_1, y_2, ..., y_n$ are independent and, for each i,

$$p(y_i^{(k)} | a, b, x_i) \propto \left[a_k N(x_i | b_k) \right]^{y_i^{(k)}}.$$

In other words,

$$p(y_i^{(k)} = 1 | a, b, x_i) = \frac{a_k N(x_i | b_k)}{p(x_i | a, b)}. \tag{7.11}$$

On the other hand, if $(a, b_1, b_2, ..., b_K)$ is subject to the probability distribution $p(a, b | x^n, y^n)$, they are independent by eq.(7.10). Hence

$$p(a, b | x^n, y^n) = p(a | x^n, y^n) \prod_{k=1}^{k} p(b_k | x^n, y^n).$$

The variable a is subject to the Dirichlet distribution with index $\alpha_k + n_k$.

$$p(a | x^n, y^n) = \frac{1}{Z} \left[\prod_{k=1}^{K} a_k^{\alpha_k + n_k - 1} \right]. \tag{7.12}$$

By using

$$H_k(b_k) = \frac{1}{2} (\frac{1}{\sigma^2} + n_k) \|b_k\|^2 - \left(\sum_i y_i^{(k)} x_i \right) b_k + Const.$$

$$= \frac{1}{2} (\frac{1}{\sigma^2} + n_k) \left\| b_k - \left(\sum_i y_i^{(k)} x_i \right) / (\frac{1}{\sigma^2} + n_k) \right\|^2 + Const.,$$

the variable b_k is subject to the normal distribution with average b_k^* and variance $(\sigma_k^*)^2$,

$$p(b_k | x^n, y^n) = N(b_k^*, (\sigma_k^*)^2), \tag{7.13}$$

where

$$b_k^* = \left(\sum_i y_i^{(k)} x_i \right) / (\frac{1}{\sigma^2} + n_k),$$

$$(\sigma_k^*)^2 = 1 / (\frac{1}{\sigma^2} + n_k).$$

Hence we obtained the Gibbs sampler from eqs.(7.11), (7.12), and (7.13).

Gibbs Sampler for Normal Mixture.

(1) A parameter set $\{(a_k, b_k); k = 1, 2, ..., K\}$ is initialized.

(2) A set of hidden variables $\{y_i = \{y_i^{(k)}\}\}$ is determined by the probability,

$$p(y_i^{(k)} = 1 | a, b, x_i) = \frac{a_k N(x_i | b_k)}{p(x_i | a, b)}.$$

(3) Using $n_k = \sum_{i=1}^{n} y_i^{(k)}$, a parameter $a = \{a_k\}$ is generated by using Dirichlet distribution,

$$p(a | x^n, y^n) = \frac{1}{Z} \Big[\prod_{k=1}^{K} a_k^{\alpha_k + n_k - 1} \Big].$$

(4) A parameter $\{b_k\}$ is generated by the normal distribution $N(b_k^*, (\sigma_k^*)^2)$ where

$$b_k^* = \Big(\sum_i y_i^{(k)} x_i \Big) / (\frac{1}{\sigma^2} + n_k),$$

$$(\sigma_k^*)^2 = 1 / (\frac{1}{\sigma^2} + n_k).$$

(5) Return to (2).

Example 52. An experiment is conducted for the case $M = 2$, $K = 2$, $\alpha_k = 1$, and $n = 100$. In Figure 7.2, for the four different true distributions, the posterior parameter sample points of $b_1 = (b_{11}, b_{12})$ and $b_2 = (b_{21}, b_{22})$ are displayed. The centers of the true distributions $(0.2B, 0.2B)$ and $(-0.2B, -0.2B)$ for $B = 4, 3, 2, 1$ are shown by the white circles. The true paramater of a is $a_0 = 0.5$. For $B \geq 3$, the posterior distributions are localized, whereas for $B \leq 2$, they are singular. Both the cross validation loss and WAIC can be applied to all cases, because both criteria can be used without normality of the posterior distribution.

7.2.2 Nonparametric Bayesian Sampler

The Gibbs sampler for mixture models can be extended as a nonparametric Bayesian sampler.

Firstly, we derive an MCMC method for a mixture model which is mathematically equivalent to the Gibbs sampler. The marginal probability density function of (b, y^n) is given by

$$p(b, y^n | x^n) = \int p(a, b, y | x^n) da.$$

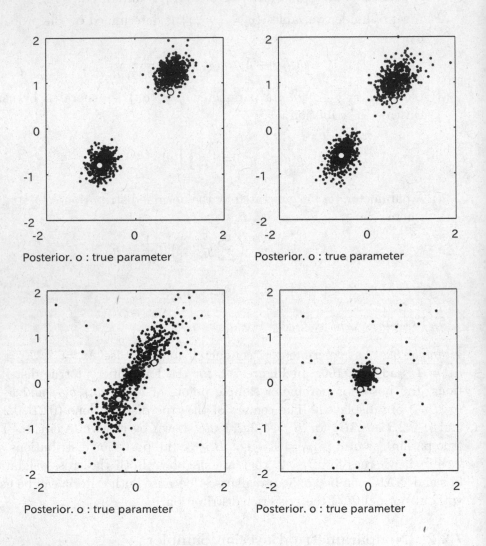

Figure 7.2: Posterior distributions of normal mixtures ($n = 100$) are displayed. The true distribution is a normal mixture which consists of two normal distributions with centers are indicated by the white circles. As the distance between two circles is made smaller, the posterior distribution becomes singular.

If we obtain the MCMC sample from $P(b, y^n | x^n)$, then the posterior distribution of $\{a\}$ or its average can be obtained by eq.(7.12). Therefore it is sufficient to make a Gibbs sampler for $p(b, y^n | x^n)$, which requires $p(b | x^n, y^n)$ and $p(y^n | b, x^n)$. The former is equal to the direct product of eq.(7.13). To derive the latter, by using

$$\int \prod_{k=1}^{K} (a_k)^{n_k} \varphi(a) da = \frac{\prod_{k=1}^{K} \Gamma(\alpha_k + n_k)}{\Gamma(n + \sum_{k=1}^{K} \alpha_k)} \frac{\Gamma(\sum_{k=1}^{K} \alpha_k)}{\prod_{k=1}^{K} \Gamma(\alpha_k)}$$

and eq.(7.9),

$$p(b, y^n | x^n) \propto \left[\frac{\prod_{k=1}^{K} \Gamma(n_k + \alpha_k)}{\Gamma(n + \sum_j \alpha_j)} \right] \varphi(b) \left[\prod_{k=1}^{K} \prod_{i=1}^{n} N(x_i | b_k)^{y_i^{(k)}} \right].$$

Hence

$$p(y^n | b, x^n) \propto \prod_{k=1}^{K} \left[\Gamma(n_k + \alpha_k) \prod_{i=1}^{n} N(x_i | b_k)^{y_i^{(k)}} \right].$$

Let $N_k(i)$ be the sum of $y_i^{(k)}$ whose sample point number is not larger than i. That is to say,

$$N_k(i) = \sum_{j=1}^{i} y_j^{(k)}.$$

Then $N_k(n) = n_k$ and

$$\Gamma(n_k + \alpha_k) = \Gamma(\alpha_k) \prod_{i=1}^{n} (\alpha_k + N_k(i) - 1)^{y_i^{(k)}}.$$

Since $\Gamma(\alpha_k)$ is a constant function of $y_i^{(k)}$,

$$p(y^n | b, x^n) \propto \prod_{i=1}^{n} \left[\prod_{k=1}^{K} \{(\alpha_k + N_k(i) - 1)N(x_i | b_k)\}^{y_i^{(k)}} \right].$$

By using

$$p(y_1) \propto \prod_{k=1}^{K} \{(\alpha_k + y_1^{(k)} - 1)N(x_i | b_k)\}^{y_1^{(k)}}, \tag{7.14}$$

$$p(y_i | y^{i-1}) \propto \prod_{k=1}^{K} \{(\alpha_k + N_k(i) - 1)N(x_i | b_k)\}^{y_i^{(k)}}. \tag{7.15}$$

The random variable $(y_1, y_2, ..., y_n)$ can be generated by iteration,

$$P(y^n|b, x^n) = p(y_1)\, p(y_2|y_1)\, p(y_3|y_1, y_2) \cdots p(y_n|y^{n-1}).$$

Note that eq.(7.15) means

$$p(y_1^{(k)} = 1) \;=\; \frac{\alpha_k N(x_1|b_k)}{\sum_{k'} \alpha_{k'} N(x_1|b_{k'})}, \tag{7.16}$$

$$p(y_i^{(k)} = 1|y^{i-1}) \;=\; \frac{(\alpha_k + N_k(i-1))N(x_i|b_k)}{\sum_{k'}(\alpha_{k'} + N_{k'}(i-1))N(x_i|b_{k'})}. \tag{7.17}$$

That is to say, y_i is determined by $N_k(i-1)$ which is the cumulative sum of $y_1, y_2, ..., y_{i-1}$. If $N_k(i-1)$ is large, then the probability that $y_i^{(k)} = 1$ is also large. This stochastic procedure is called "the Chinese restaurant process", where i is a guest of a resaurant and k is the number of a table. The ith guest determines a table according to the numbers of persons sitting at tables.

If $K \to \infty$ and $\alpha_k = \alpha/K$, then this Gibbs sampler determined by eqs.(7.13), (7.16), and (7.17) gives the statistical estimation of the nonparametric Bayesian method. We obtained the following algorithm.

Nonparametric Bayesian Sampler for Normal Mixture.

(1) A parameter set $\{(a_k, b_k); k = 1, 2, ..., K\}$ is initialized.

(2) A set of hidden variables $\{y_i = \{y_i^{(k)}\}\}$ is iteratively determined by the probability,

$$p(y_1^{(k)} = 1) \;=\; \frac{\alpha_k N(x_1|b_k)}{\sum_k \alpha_k N(x_1|b_k)},$$

$$p(y_i^{(k)} = 1|y^{i-1}) \;=\; \frac{(\alpha_k + N_k(i-1))N(x_i|b_k)}{\sum_k (\alpha_k + N_k(i-1))N(x_i|b_k)},$$

where $N_k(i) = \sum_{j=1}^{i} y_j^{(k)}$.

(3) A parameter $\{b_k\}$ is generated by the normal distribution $N(b_k^*, (\sigma_k^*)^2)$ where

$$b_k^* \;=\; \Big(\sum_i y_i^{(k)} x_i\Big) / (\frac{1}{\sigma^2} + n_k),$$

$$(\sigma_k^*)^2 \;=\; 1/(\frac{1}{\sigma^2} + n_k).$$

(4) Return to (2).

Remark 57. (1) This algorithm is a Gibbs sampler for (b, y^n), whereas the previous one was applied to (a, b, y^n). The expectation operation over a is analytically performed.

(2) In statistics, a nonparametric estimation of a density function of X for a given X^n is usually defined by

$$\hat{p}(x) = \frac{1}{n} \sum_{i=1}^{n} \rho\left(\frac{x - X_i}{n^\alpha}\right),$$

where $\rho(x)$ is some kernel function such as a normal distribution, and α is an optimized controlling parameter. In this method, the estimated density is a mixture of n functions. The nonparametric Bayesian method is formally defined by the mixture of the infinite number of functions, however, it requires the very small $\{\alpha_k = \alpha/K\}$, so that the number K essentially used in MCMC is finite. The optimal hyerparameter α that minimizes the generalization loss can be evaluated by the cross validation and WAIC. In order to ensure the generalization loss is smaller, infinite components should be controlled close to zero, therefore the prior effect should be made stronger. The generalization error by the mixture model with the appropriate finite number of components is smaller than that from the mixture of infinite components.

7.3 Numerical Approximation of Observables

By using the Markov chain Monte Carlo method, we can numerically calculate Bayesian observables.

7.3.1 Generalization and Cross Validation Losses

Let $\{w_k; k = 1, 2, ..., K\}$ be a set of posterior parameters. The generalization loss is numerically approximated by

$$G_n = -\frac{1}{T} \sum_{t=1}^{T} \log\left(\frac{1}{K} \sum_{k=1}^{K} \frac{1}{p(X_t|w_k)}\right),$$

where $\{X_t\}$ is a set of random variables which are independent of the sample used in the posterior distribution. In general T should be very large, $T >> n$, in order to minimize the fluctuation. ISCV indexISCV and WAIC can also be approximated numerically.

$$\text{ISCV} = \frac{1}{n} \sum_{i=1}^{n} \log \Big(\frac{1}{K} \sum_{k=1}^{K} \frac{1}{p(X_i|w_k)} \Big),$$

$$\text{WAIC} = -\frac{1}{n} \sum_{i=1}^{n} \log \Big(\frac{1}{K} \sum_{k=1}^{K} p(X_i|w_k) \Big) + V_n,$$

$$V_n = \frac{1}{n} \sum_{i=1}^{n} \Big[\frac{1}{K} \sum_{k=1}^{K} \Big(\log p(X_i|w_k) \Big)^2 - \Big(\frac{1}{K} \sum_{k=1}^{K} \log p(X_i|w_k) \Big)^2 \Big],$$

where, in calculation of V_n, $(K/(K-1))V_n$ is more appropriate than V_n, because it estimates the sum of the variances of $\log p(X_i|w_k)$ over the posterior distribution.

Remark 58. In order to calculate the above observables, we need

$$\{p(X_i|w_k); i = 1, 2, ..., n, \ \ k = 1, 2, ..., K\}.$$

Several softwares have parallel computation architectures. In such a case, for every parameter w_k, the set

$$p(X_1|w_k), \ \ p(X_2|w_k), \ \ \cdots \ \ p(X_n|w_k)$$

can be simultaneously calculated. Also for every X_i, the set

$$p(X_i|w_1), \ \ p(X_i|w_2), \ \ \cdots \ \ p(X_i|w_K)$$

can be simultaneously calculated. Once $\{p(X_i|w_k)\}$ is obtained, then the above computation is not so heavy in general. For neural networks and normal mixtures, this method is recommended for reducing the computational costs.

7.3.2   Numerical Free Energy

Even if the posterior parameters $\{w_k\}$ are obtained, it is not enough to numerically estimate the free energy or the minus log marginal likelihood. Here we study a method to calculate them.

Definition 24. Let $\beta > 0$. The average of an arbitrary function $f(w)$ over the generalized posterior distribution is defined by

$$\mathbb{E}_w^{(\beta)}[f(w)] = \frac{\displaystyle\int f(w)\varphi(w) \prod_{i=1}^{n} p(X_i|w)^\beta dw}{\displaystyle\int \varphi(w) \prod_{i=1}^{n} p(X_i|w)^\beta dw}.$$

Then the case $\beta = 1$ is equal to the posterior average, $\mathbb{E}_w^{(1)}[\ ] = \mathbb{E}_w[\ ]$.

Theorem 23. *By using the minus log likelihood function,*

$$L_n(w) = -\frac{1}{n} \sum_{i=1}^{n} \log p(X_i|w).$$

The free energy or the minus log marginal likelihood is given by

$$F_n = \int_0^1 \mathbb{E}_w^{(\beta)}[nL_n(w)]d\beta.$$

Proof. Let us define a function $F(\beta)$

$$F_n(\beta) = -\log \int \varphi(w) \prod_{i=1}^{n} p(X_i|w)^\beta dw.$$

Then $F_n(0) = 0$ and

$$F_n = F_n(1).$$

By using,

$$\frac{\partial}{\partial \beta} \Big(\prod_{i=1}^{n} p(X_i|w) \Big)^\beta = \log \Big(\prod_{i=1}^{n} p(X_i|w) \Big) \Big(\prod_{i=1}^{n} p(X_i|w) \Big)^\beta,$$

it follows that

$$
\begin{aligned}
F_n(1) &= \int_0^1 d\beta\, \frac{\partial F}{\partial \beta}(\beta) \\
&= \int_0^1 d\beta\, \frac{\displaystyle\int (nL_n(w))\, \varphi(w) \prod_{i=1}^{n} p(X_i|w)^\beta dw}{\displaystyle\int \varphi(w) \prod_{i=1}^{n} p(X_i|w)^\beta dw},
\end{aligned}
$$

which shows the theorem. $\qquad\square$

A calculation method of the free energy is derived by the same method as the above theorem.

Let $\{\beta_k; k = 0, 1, ..., J\}$ be a sequence,

$$0 = \beta_0 < \beta_1 < \cdots < \beta_J = 1.$$

Since $Z_n(0) = 1$,

$$
\begin{aligned}
Z_n(1) &= \prod_{j=0}^{J-1} \left(\frac{Z_n(\beta_{k+1})}{Z_n(\beta_k)} \right) \\
&= \prod_{j=0}^{J-1} E_w^{(\beta_k)} \left[e^{-(\beta_{k+1}-\beta_k)nL_n(w)} \right],
\end{aligned} \tag{7.18}
$$

The free energy is given by

$$
\begin{aligned}
F_n(1) &= -\log Z_n(1) \\
&= -\sum_{j=0}^{J-1} \log E_w^{(\beta_k)} \left[e^{-(\beta_{k+1}-\beta_k)nL_n(w)} \right].
\end{aligned} \tag{7.19}
$$

Here we need the posterior distribution for $\beta_1, ..., \beta_{J-1}$

$$
E^{(\beta_1)}[\ ], E^{(\beta_2)}[\ ], ..., E^{(\beta_{J-1})}[\ ]
$$

which can be obtained by the parallel tempering.

Remark 59. This method sometimes involves heavy computational costs. If the posterior distribution can be approximated by some normal distribution, then eq.(4.57) can be applied, otherwise WBIC can be employed, however, the difference between the free energy and WBIC is $\log \log n$ or constant order. If minimizing the free energy according to a hyperparameter, then the derivative of F_n by the hyperparameter can be calculated for the smaller computational cost.

Remark 60. By using a probability density function,

$$
p(w) = \frac{1}{Z} \varphi(w) \exp(-nL_n(w)),
$$

the expectation $\mathbb{E}_w[\ ]$ is defined by $p(w)$, Then

$$
Z = \frac{1}{\mathbb{E}_w[\exp(nL_n(w))]}.
$$

However, this method is not appropriate for calculating Z because $\exp(nL_n(w))$ takes the large values at the small $p(w)$.

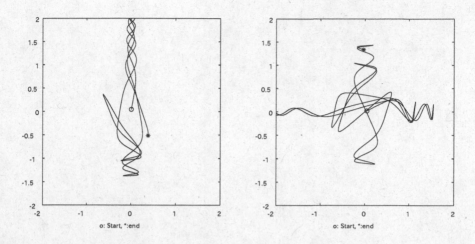

Figure 7.3: Examples of trajectories by Hamiltonian equation. The trajectories depend on initial conditions. The label 'o' and '*' show the start and end points respectively.

7.4 Problems

1. Let us study a probability density on $W = [-2, 2] \times [-1, 1]$ defined by

$$p(u, v) \propto \exp(-nv^2(u+1)^2(u-1)^4).$$

Then the set of all points which attain the maximum of $p(u, v)$ is

$$W_0 = \{(u, v) \in W \ \ v(u+1)(u-1) = 0\}.$$

Prove that, when $n \to \infty$, this distribution coverges to $\delta(u-1)\delta(v)$. Explain why it does not coverge to the all neighborhoods of W_0.

2. For the same Hamiltonian function used in Example 51, the trajectories of Hamiltonian equation are shown in Figure 7.3. Explain the reason why Hamiltonian Monte Carlo can cover the entire parameter set.

3. In Example 52, the posterior distributions of the parameter a are not displayed. For each case, explain the shape of the posterior distribution of a. Also answer whether it is localized or not.

Chapter 8

Information Criteria

In the foregoing chapters, we derived the theoretical behaviors of Bayesian observables for a given set of a true distribution, a statistical model, and a prior, $(q(x), p(x|w), \varphi(w))$. In the real world, we do not know the true distribution $q(x)$, hence we need methods to estimate observables without any information about $q(x)$. Information criteria are made to overcome such problems. In this section we explain several information criteria from the two viewpoints, model selection and hyperparameter optimization. In each viewpoint, the properties of the generalization loss and the free energy or the minus log marginal likelihood are investigated. This chapter consists of the following contents.

- Model Selection

    - Generalization Loss: CV, AIC, TIC, DIC, WAIC
    - Free Energy: F, BIC, WBIC

- Hyperparameter optimization

    - Generalization Loss: CV, WAIC
    - Free Energy: F, DF

8.1 Model Selection

In this section we study a model selection problem. When we have several candidate models and need to select one of them, the model selection problem occurs. There are two methods in model selection, minimizing the

231

generalization loss and the free energy. The aim of minimizing the generalization loss is equivalent to minimizaing the Kullback-Leibler distance from the true density and the predictive one, whereas that of the minimizing the free energy is to maximize the posterior probability of a statistical model and a prior for a given set of data.

8.1.1   Criteria for Generalization Loss

Let us introduce the definitions of several information criteria which are used for estimation of the generalization loss. Since the generalization losses for Bayesian, maximum likelihood, maximum *a posteriori*, and posterior mean methods are different, we have to understand which generalization loss an information criterion estimates.

Remark 61. In this book, the information criteria are defined as estimators of the generalization loss

$$-\mathbb{E}_X[\log \hat{p}(X)], \tag{8.1}$$

where $\hat{p}(x)$ is the estimated probability density of x by a statistical estimation method. Since the original Akaike information criterion AIC was defined to estimate

$$-2n \times \mathbb{E}_X[\log \hat{p}(X)], \tag{8.2}$$

resulting that many information criteria were normalized so that they estimate the same scale loss as AIC. If one needs information criteria which have the same scale loss as AIC, $2n$ times values shown in this book should be used. If eq.(8.1) is used, then the difference between candidate models is measured by the scale according to the Kullback-Leibler distance, whereas, if eq.(8.2) is used, then it is measured by the scale according to the number of parameters.

Definition 25. The leave-one-out cross validation criterion CV and the importance sampling cross validation criterion ISCV are respectively defined by

$$\mathrm{CV} = -\frac{1}{n}\sum_{i=1}^{n} \log \mathbb{E}_w^{(-i)}[p(X_i|w)], \tag{8.3}$$

$$\mathrm{ISCV} = \frac{1}{n}\sum_{i=1}^{n} \log \mathbb{E}_w[1/p(X_i|w)], \tag{8.4}$$

where $\mathbb{E}_w[\ ]$ and $\mathbb{E}_w^{(-i)}[\ ]$ show the ordinary posterior average and the posterior average leaving X_i out, respectively. Both criteria estimate the Bayesian generalization loss, if $\{X_i\}$ are independent. If the posterior distributions in the above definitions are exactly realized and the averages are finite, then CV = ISCV. However, if they are numerically approximated, for example by Markov Chain Monte Carlo, then CV $\neq$ ISCV. In order to calculate CV, all posterior distributions using $X^n \setminus X_i$ for $i = 1, 2, ...n$ are necessary, whereas ISCV can be calculated by one posterior distribution using X^n.

Definition 26. The Akaike information criteria (AIC) and that for Bayes (AIC$_b$) are defined for the maximum likelihood and Bayesian methods respectively,

$$\text{AIC} = -\frac{1}{n}\sum_{i=1}^{n} \log p(X_i|\hat{w}) + \frac{d}{n}, \qquad (8.5)$$

$$\text{AIC}_b = -\frac{1}{n}\sum_{i=1}^{n} \log \mathbb{E}_w[p(X_i|w)] + \frac{d}{n}, \qquad (8.6)$$

where $\hat{w}$ is the maximum likelihood estimator. The Takeuchi information criteria (TIC) and that for Bayes (TIC$_b$) are defined for the maximum likelihood and Bayesian methods, respectively,

$$\text{TIC} = -\frac{1}{n}\sum_{i=1}^{n} \log p(X_i|\hat{w}) + \frac{1}{n}\text{tr}(I(\hat{w})J(\hat{w})^{-1}), \qquad (8.7)$$

$$\text{TIC}_b = -\frac{1}{n}\sum_{i=1}^{n} \log \mathbb{E}_w[p(X_i|w)] + \frac{1}{n}\text{tr}(I(\overline{w})J(\overline{w})^{-1}), \qquad (8.8)$$

where $\overline{w} = \mathbb{E}_w[w]$ and

$$I(w) = \frac{1}{n}\sum_{i=1}^{n} \nabla \log p(X_i|w)(\nabla \log p(X_i|w))^T, \qquad (8.9)$$

$$J(w) = -\frac{1}{n}\sum_{i=1}^{n} \nabla^2 \log p(X_i|w). \qquad (8.10)$$

Note that the original AIC and TIC are criteria for the generalization loss of the maximum likelihood method, whereas AIC$_b$ and TIC$_b$ are their modifications for Bayesian estimation. In general, the generalization loss of Bayes is different from that of the maximum likelihood method and AIC $\neq$ AIC$_b$ and TIC $\neq$ TIC$_b$.

Definition 27. The deviance information criterion (DIC) is defined by

$$\text{DIC} \;=\; \frac{1}{n}\sum_{i=1}^{n}\log p(X_i|\overline{w}) - \frac{2}{n}\sum_{i=1}^{n}\mathbb{E}_w[\log p(X_i|w)], \qquad (8.11)$$

where $\overline{w} = \mathbb{E}_w[w]$. It seems that DIC is made for estimating the generalization loss of $p(x|\overline{w})$ rather than the predictive distribution $\mathbb{E}_w[p(x|w)]$. However, it might be employed for Bayesian or the maximum *a posteriori* method. The widely applicable information criterion (WAIC) is defined by

$$\text{WAIC} \;=\; -\frac{1}{n}\sum_{i=1}^{n}\log \mathbb{E}_w[p(X_i|w)] + \frac{1}{n}\sum_{i=1}^{n}\mathbb{V}_w[\log p(X_i|w)], \quad (8.12)$$

which estimates the generalization loss of the predicitive density.

The behaviors of information criteria depend on the condition of a true distribution $q(x)$, a statistical model $p(x|w)$, and a prior $\varphi(w)$. Let us consider (A) a regular and realizable case, (B) a egular and unrealizable case, and (C) a nonregular case.

(A) Regular and Realizable Case

If a true distribution is realizable by and regular for a statistical model, and if the posterior distribution can be approximated by some normal distribution, the generalization and training losses by Bayes, maximum *a posteriori*, posterior mean, and the maximum likelihood methods have the same asymptotic expansion as

$$\mathbb{E}[G_n] \;=\; L(w_0) + \frac{d}{2n} + o(1/n),$$

$$\mathbb{E}[T_n] \;=\; L(w_0) - \frac{d}{2n} + o(1/n),$$

where d is the dimension of the parameter. In this case an arbitrary criterion of CV, ISCV, AIC, AIC$_b$, TIC, TIC$_b$, DIC, and WAIC satisfies

$$\mathbb{E}[\text{Criterion}] \;=\; L(w_0) + \frac{d}{2n} + o(1/n). \qquad (8.13)$$

Hence arbitrary information criteria can be employed. The asymptotic standard deviations of all criteria are also equal to each other.

(B) Regular and Unrealizable Case

If a true distribution is regular for but unrealizable by a statistical model, we define

$$\nu_0 = \frac{1}{2}\operatorname{tr}(I_0 J_0^{-1}),$$

where

$$
\begin{aligned}
I_0 &= \int \nabla \log p(X_i|w_0)(\nabla \log p(X_i|w_0)^T dx, \\
J_0 &= -\int \nabla^2 \log p(X_i|w_0) dx.
\end{aligned}
$$

If a true distribution is realizable by a statistical model, then $\nu_0 = d/2$. If the posterior distribution can be approximated by some normal distribution, then the average generalization and training losses of Bayes and the maximum likelihood methods are

$$
\begin{aligned}
\mathbb{E}[G_n] &= L(w_0) + \frac{d}{2n} + o(1/n), & (8.14) \\
\mathbb{E}[T_n] &= L(w_0) + \frac{d - 4\nu_0}{2n} + o(1/n), & (8.15) \\
\mathbb{E}[G_n(ML)] &= L(w_0) + \frac{\nu_0}{n} + o(1/n), & (8.16) \\
\mathbb{E}[T_n(ML)] &= L(w_0) - \frac{\nu_0}{n} + o(1/n), & (8.17)
\end{aligned}
$$

where $G_n(ML)$ and $T_n(ML)$ are the generalization and training losses of the maximum likelihood method, respectively. The generalization and training losses of the maximum *a posteriori* and posterior mean methods are asymptotically equal to those of the maximum likelihood.

The averages of the cross validations are equal to the generalization loss asymptotically,

$$
\begin{aligned}
\mathbb{E}[CV] &= L(w_0) + \frac{d}{2n} + o(1/n), & (8.18) \\
\mathbb{E}[ISCV] &= L(w_0) + \frac{d}{2n} + o(1/n). & (8.19)
\end{aligned}
$$

Since $\mathbb{E}[T_n] = L(w_0) + (d - 4\nu_0)/(2n)$,

$$
\begin{aligned}
\mathbb{E}[AIC] &= L(w_0) + \frac{d - \nu_0}{n} + o(1/n), & (8.20) \\
\mathbb{E}[AIC_b] &= L(w_0) + \frac{3d - 4\nu_0}{2n} + o(1/n). & (8.21)
\end{aligned}
$$

By the definitions of TIC and TIC_b,

$$\mathbb{E}[\text{TIC}] = L(w_0) + \frac{\nu_0}{2n} + o(1/n), \tag{8.22}$$

$$\mathbb{E}[\text{TIC}_b] = L(w_0) + \frac{d}{2n} + o(1/n). \tag{8.23}$$

By using eq.(4.55) and eq.(4.56),

$$\mathbb{E}[\text{DIC}] = L(w_0) + \frac{3d - 4\nu_0}{2n} + o(1/n), \tag{8.24}$$

$$\mathbb{E}[\text{WAIC}] = L(w_0) + \frac{d}{2n} + o(1/n). \tag{8.25}$$

Therefore, CV, ISCV, TIC_b, and WAIC can be used for estimating the Bayesian generalization loss. To estimate the generalization loss of the maximum likelihood method, TIC is available.

(C) Nonregular Case

If a true distribution is not regular for a statistical model, let λ and ν be the real log canonical threshold and a singular fluctuation defined by

$$\lambda = \text{Real Log Canonical Threshold}, \tag{8.26}$$

$$\nu = \frac{1}{2}\mathbb{E}_\xi[\text{Fluc}(\xi)]. \tag{8.27}$$

Also let μ be a constant defined by Theorem 19. Then

$$\mathbb{E}[G_n] = L(w_0) + \frac{\lambda}{n} + o(1/n), \tag{8.28}$$

$$\mathbb{E}[T_n] = L(w_0) + \frac{\lambda - 2\nu}{n} + o(1/n), \tag{8.29}$$

$$\mathbb{E}[G_n(ML)] = L(w_0) + \frac{\mu}{n} + o(1/n), \tag{8.30}$$

$$\mathbb{E}[T_n(ML)] = L(w_0) - \frac{\mu}{n} + o(1/n). \tag{8.31}$$

In general, $\mu \gg \lambda$, hence Bayesian estimation attains the smaller generalization loss than the maximum likelihood, maximum *a posteriori*, and posterior mean methods. In this case, the average cross validation loss is equal to the generalization loss,

$$\mathbb{E}[\text{CV}] = L(w_0) + \frac{\lambda}{n} + o(1/n), \tag{8.32}$$

$$\mathbb{E}[\text{ISCV}] = L(w_0) + \frac{\lambda}{n} + o(1/n). \tag{8.33}$$

By the definition of AIC and AIC_b,

$$\mathbb{E}[AIC] \quad = \quad L(w_0) + \frac{d - \mu}{n} + o(1/n), \tag{8.34}$$

$$\mathbb{E}[AIC_b] \quad = \quad L(w_0) + \frac{\lambda - 2\nu + d}{n} + o(1/n). \tag{8.35}$$

Since TIC and TIC_b are undefined because $J(w)$ is not invertible,

$$\mathbb{E}[TIC] \quad = \quad \text{Undefined}, \tag{8.36}$$

$$\mathbb{E}[TIC_b] \quad = \quad \text{Undefined}. \tag{8.37}$$

The posterior average parameter $\mathbb{E}_w[w]$ is not in the neighborhood of the optimal parameter set, hence there exists $C > 0$ such that

$$\mathbb{E}[DIC] \quad = \quad L(w_0) + C + o(1), \tag{8.38}$$

$$\mathbb{E}[WAIC] \quad = \quad L(w_0) + \frac{\lambda}{n} + o(1/n). \tag{8.39}$$

In this case, the Bayes generalization loss is estimated by the cross validation and WAIC. Note that in nonregular cases, any information criterion is not yet known which can estimate the generalization loss of the maximum likelihood, maximum *a posteriori*, and posterior mean methods, because the posterior distribution is far from any normal distribution. It seems that the constant μ cannot be estimated because it depends on the optimal parameter w_0.

Remark 62. The above results hold for the assumption that a sample X^n consists of independent sample points. For a case when $\{Y^n\}$ is conditionally independent for a given x^n, then information criteria can estimate the generalization loss if it can in independent cases, whereas the cross validation criterion cannot. For example, the cross validation loss cannot be employed in a linear prediction of time series, whereas information criteria can be.

Remark 63. From the mathematical point of view, the information criteria need asymptotic condition $n \to \infty$. In fact, AIC, AIC_b, TIC, TIC_b, and DC require the asymptotic normality, resulting that the sample size n should be large enough. However, WAIC does not require the asymptotic normality, hence it can estimate the generalization loss even if n is not so large. In many singular statistical models, WAIC can estimate the generalization loss even if n is small experimentally.

Example 53. Let $x \in \mathbb{R}^2$ and $N(x)$ be a normal distribution on $\mathbb{R}^2$ whose average is zero and covariance matrix is the 2×2 identity matrix.

$$N(x) = \frac{1}{2\pi} \exp(-\frac{1}{2}\|x\|^2).$$

We study a statistical model

$$p(x|a, b, c) = aN(x - b) + (1 - a)N(x - c),$$

where $0 \le a \le 1$ and $b, c \in \mathbb{R}^2$ are parameters. For a prior of a, we adopt a Dirichlet distribution,

$$\varphi_1(a) \propto (a(1 - a))^{\alpha - 1},$$

where $\alpha > 0$ is a hyperparameter. For a prior of (b, c),

$$\varphi_2(b, c) \propto \exp(-\frac{\|b\|^2 + \|c\|^2}{2B^2}),$$

where B is a hyperparameter. Several cases are studied experimentally.
(1) Regular and realizable case.

$$q(x) = p(x|a_0, b_0, c_0),$$

where $a_0 = 0.5$, $b_0 = (2, 2)$, and $c = (-2, -2)$.
(2) Regular and unrealizable case.

$$q(x) = a_0 N((x - b_0)/\sigma)/\sigma + (1 - a_0)N((x - c_0)/\sigma)/\sigma,$$

where $a_0 = 0.5$, $\sigma = 0.8$, $b_0 = (2, 2)$, and $c_0 = (-2, -2)$.
(3) Nonregular and realizable case.

$$q(x) = p(x|a_0, b_0, c_0),$$

where $a_0 = 0$, $b_0 = (0, 0)$, and $c = (0, 0)$.
(4) Nonregular and unrealizable case.

$$\dot{q}(x) = a_0 N((x - b_0)/\sigma)/\sigma + (1 - a_0)N((x - c_0)/\sigma)/\sigma,$$

where $a_0 = 0$, $\sigma = 0.8$, $b_0 = (0, 0)$, and $c = (0, 0)$.
(5) Delicate case.

$$q(x) = a_0 N((x - b_0)/\sigma)/\sigma + (1 - a_0)N((x - c_0)/\sigma)/\sigma,$$

Cases		G	ISCV	AIC$_b$	DIC	WAIC
(1) Regular	Ave	0.0254	0.0254	0.0251	0.0247	0.0254
Realizable	Std	0.0169	0.0162	0.0164	0.0162	0.0162
(2) Regular	Ave	0.1089	0.1043	0.1184	0.1110	0.1043
Unrealizable	Std	0.0124	0.0362	0.0381	0.0372	0.0362
(3) Nonreg.	Ave	0.0129	0.0160	0.0418	0.0034	0.0158
Realizable	Std	0.0085	0.0088	0.0118	0.0283	0.0088
(4) Nonreg.	Ave	0.0983	0.1036	0.1409	0.1049	0.1036
Unreal.	Std	0.0067	0.0399	0.0412	0.0455	0.0399
(5) Delicate	Ave	0.0384	0.0384	0.0479	0.0001	0.0387
	Std	0.0175	0.0239	0.0232	0.0537	0.0241
(6) Unbal.	Ave	0.0276	0.0255	0.0343	-0.1618	0.0225
	Std	0.0169	0.0267	0.0156	0.3568	0.0235

Table 8.1: Experimental results in Example 53. In the table, averages and standard deviations of normalized values G-S, ISCV-S_n, AIC$_b$ -S_n, DIC -S_n, and WAIC-S_n are displayed.

where $a_0 = 0.5$, $\sigma = 0.95$, $b_0 = (0.5, 0.5)$, and $c = (-0.5, -0.5)$.
(6) Unbalanced case.

$$q(x) = a_0 N(x - b_0) + (1 - a_0) N(x - c_0),$$

where $a_0 = 0.01$, $b_0 = (2, 2)$, and $c = (-2, -2)$.
In each case, the average and empirical entropies of the true distributions are defined by

$$S_0 = - \int q(x) \log q(x) dx,$$

$$S_{0n} = -\frac{1}{n} \sum_{i=1}^{n} \log q(X_i).$$

For the case $n = 100$, the posterior distributions were built by the Gibbs sampler, in which the burn-in was 200 and the number of posterior parameters were 1000. Hyperparameters were set as $\alpha = 0.5$ and $B = 10$. We conducted 100 independent experiments for each condition. In Table 53, 'Ave' and 'Std' show their averages and standard deviations of $G_n - S$, ISCV $- S_n$, AIC$_b - S_n$, DIC $- S_n$, and WAIC$-S_n$.
(1) If a true distribution was realizable by and regular for a statistical model, then all information criteria estimated the generalization loss well.

(2) If a true distribution was unrealizable by and regular for a statistical model, then AIC overestimated the generalization loss.

(3) through (5) In nonregular and delicate cases, ISCV and WAIC were more accurate than AIC and DIC.

(6) In an unbalanced case, ISCV, WAIC, and AIC were more accurate than DIC. Note that a few sample points were generated from the first component. As an unbiased estimator, ISCV was better than WAIC and AIC, however, the variance of ISCV was larger than WAIC and AIC. AIC had the smallest variance. Their intervals $[m - 2\sigma, m + 2\sigma]$ where m and σ are averages and standard deviations were

$$
\begin{aligned}
G - S &: \quad [-0.0062, 0.614] \\
\text{ISCV} - S_n &: \quad [-0.0279, 0.0789] \\
\text{AIC} - S_n &: \quad [0.0031, 0.0655] \\
\text{DIC} - S_n &: \quad [-0.8754, 0.5518] \\
\text{WAIC} - S_n &: \quad [-0.0245, 0.0695]
\end{aligned}
$$

Therefore, not only the cross validation loss but also information criteria contain important information.

Example 54. Examples of model selections are shown in sections 2.4 and 2.5. If a statistical model has hierarchical structure or hidden variables, then the posterior distribution cannot be approximated by any normal distribution in general, hence we can apply the cross validation and WAIC, but not AIC or DIC. If we need a neural network with many hidden units or a normal mixture with many components, the MCMC process sometimes fails because of local minima. If such models have a few redundant hidden parts, the MCMC rather easily attains the posterior distribution. If Bayesian estimation is applied to such statistical models, the generalization losses do not increase much, hence we recommend a model which has a few redundant parts.

8.1.2   Comparison of ISCV with WAIC

In typical experiments, ISCV is almost equal to WAIC. First, we show that if a sample consists of independent random variables, then ISCV and WAIC are asymptotically equivalent as random variables. Let $\mathcal{T}_n(\alpha)$ be a function defined in eq.(3.11) in Definition 8.

Theorem 24. *Assume that $X_1, X_2, ..., X_n$ are independent and that*

$$
\sup_{|\alpha| \leq 1} \left| \left(\frac{d}{d\alpha} \right)^4 \mathcal{T}_n(\alpha) \right| = O_p\left(\frac{1}{n^2}\right). \tag{8.40}
$$

Then the following equation holds,

$$\text{ISCV} = \text{WAIC} + O_p(\frac{1}{n^2}). \tag{8.41}$$

Proof. ISCV is defined by

$$\text{ISCV} = \frac{1}{n} \sum_{i=1}^{n} \log \mathbb{E}_w[\, 1/p(X_i|w)\,].$$

By the definition of $\mathcal{T}_n(\alpha)$ in eq.(3.11)

$$\text{ISCV} = \mathcal{T}_n(-1).$$

By using the mean value theorem, there exists $|\beta^*| < 1$ such that

$$\mathcal{T}_n(-1) \;=\; -\mathcal{T}_n'(0) + \frac{1}{2}\mathcal{T}_n''(0) - \frac{1}{6}\mathcal{T}_n^{(3)}(0) + \frac{1}{24}\mathcal{T}_n^{(4)}(\beta^*).$$

On the other hand WAIC is defined by

$$\text{WAIC} \;=\; T_n + V_n = -\mathcal{T}_n(1) + \mathcal{T}_n''(0).$$

By using the mean value theorem, there exists $|\beta^{**}| < 1$ such that

$$-\mathcal{T}_n(1) \;=\; -\mathcal{T}_n'(0) - \frac{1}{2}\mathcal{T}_n''(0) - \frac{1}{6}\mathcal{T}_n^{(3)}(0) - \frac{1}{24}\mathcal{T}_n^{(4)}(\beta^{**}).$$

Hence

$$\text{WAIC} = -\mathcal{T}_n'(0) + \frac{1}{2}\mathcal{T}_n''(0) - \frac{1}{6}\mathcal{T}_n^{(3)}(0) - \frac{1}{24}\mathcal{T}_n^{(4)}(\beta^{**}),$$

which completes the theorem. $\qquad\qquad\qquad\qquad\qquad\qquad\qquad\square$

Remark 64. (1) This theorem holds, even if the posterior distribution cannot be approximated by any normal distribution. By the proof, it is also derived that

$$\text{ISCV} = \text{WAIC} + \frac{1}{12}\mathcal{T}_n^{(4)}(0) + o_p(n^{-2}).$$

By Theorem 26, if the posterior distribution can be approximated by some normal distribution, then $\mathcal{T}_n^{(4)}(0) = o_p(n^{-2})$, resulting that the difference between ISCV and WAIC is smaller than $O_p(n^{-2})$.

(2) Assume that there exist constants g_1 and g_2 which satisfy

$$\mathbb{E}[G_n] \;=\; g_1 + \frac{g_2}{n} + o(1/n).$$

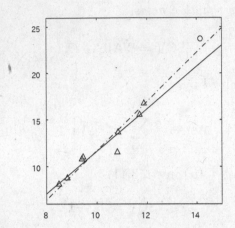

Figure 8.1:   The horizontal and vertical lines show the pairs of $(\log(\text{radius}), \log(\text{mass}))$ in the solar system. The circle corresponds to the sun, which is the leverage sample point. In fact, the regression line estimated by including the sun is given by the dotted line, whereas regression by not including it is shown by the solid line.

By the definition

$$\mathbb{E}[\text{CV}] = \mathbb{E}[G_{n-1}].$$

Hence

$$\mathbb{E}[\text{CV}] - \mathbb{E}[G_n] = \mathbb{E}[G_{n-1}] - \mathbb{E}[G_n] = O(1/n^2).$$

By the above theorem

$$\mathbb{E}[\text{WAIC}] - \mathbb{E}[G_n] = O(1/n^2).$$

Therefore, CV and WAIC have asymptotically the same approximators of the generalization loss, if a sample consists of independent random variables.

Remark 65. (Comparison of ISCV and WAIC) In the numerical experiments, the difference between ISCV and WAIC is very small in many cases, however, sometimes they are different. First, if a sample $\{(X_i, Y_i)\}$ is dependent, then the averages of CV and ISCV are different from that of the generalization loss. On the other hand, the averages of WAIC are asymptotically equal to those of the generalization loss if a sample consists of conditionally independent variables. Second, in statistical estimation of the conditional probability $q(y|x)$, if n is not enough large to ensure

$$q(x) \approx \frac{1}{n} \sum_{i=1}^{n} \delta(x - X_i),$$

then ISCV is different from WAIC. Thirdly, if a sample contains a leverage sample point, then variance of ISCV diverges which is different from WAIC. The third sample can be understood as a special case of the second one. If a leverage sample point is contained in a sample, then the data analyst should reconsider whether such a point should be included in a sample. A leverage sample point can be found by the following procedure. If ISCV is not equal to WAIC, then for every sample point X_i, the partial functional variance

$$\mathbb{V}_w[\log p(Y_i|X_i, w)]$$

is calculated. If it is larger than the others, then X_i is a leverage sample point.

Remark 66. The importance sampling cross validation loss diverges if a leverage sample point is contained [57] [20]. Recently, a new method for numerical approximation of the cross validation was devised in which the posterior distribution is replaced by the Pareto distribution [76]. This method gives the approximation of the cross validaiton loss. WAIC is not an approximation of the cross validation loss but is an estimator of the generalization loss. If a sample is dependent, then the cross validation loss is not an estimator of the generalization loss.

Example 55. (Leverage sample point) Let $\{X_i\}$ be the $\{\log(\text{radius})\}$ of stars in the solar system, Mercury, Venus, Earth,..., and $\{Y_i\}$ be $\{\log(\text{mass})\}$. If we study a simple regression problem, $Y = aX + b + noise$, then the datum of the sun is a leverage sample point. In Figure 8.1, the circle shows the datum of the sun. A regression line without the sun is shown by the solid line whereas regression with the sun is shown by the dotted line. The (X, Y) of the sun may not be estimated from other data, hence the cross validation fails. Even in such a case, information criteria AIC and WAIC can be used to estimate the statistical estimation error.

Example 56. (Classification problem) Let us study a classification problem $q(z|x, y)$ using a neural network, where $(x, y) \in \mathbb{R}^2$ and the true output z is set by a function,

$$z = \begin{cases} 1 & y > \sin(\pi x/2) \\ 0 & \text{otherwise} \end{cases}.$$

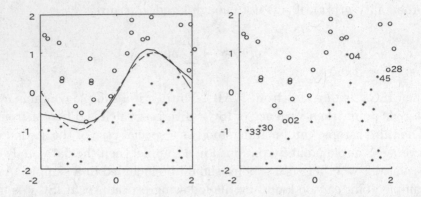

Figure 8.2: Classification problem. Two categories in two dimensinal space are classified by a neural network. The letters 'o' and '*' show sample points classified as one and zero. The solid and dotted lines show the estimated and true boundaries. The sample points near the boundary are leverage sample points.

A sample of 50 points is shown in Figure 8.2. The solid and dotted lines show the estimated and true classification boundary respectively. The letters 'o' and '*' are sample points classified as one and zero by the true rule respectively. A three-layered neural network which has input units $M = 2$, hidden units $H = 5$, and an output unit $N = 1$ was employed for learn the classification rule. The posterior distribution is approximated by the Metropolis method explained in the previous chapter. In the classification problem, sample points near the boundary strongly affect the statistical inference: in fact, the classification result for such a point is not estimated from the other sample points. In Figure 8.2, several points which are displayed with numbers are leverage sample points. The partial functional variance of the ith sample point

$$V_i = \mathbb{V}_w[\log p(X_i|w)]$$

shows the strength of the sample point's effect. Figure 8.3 shows such $\{V_i\}$ for each i. The larger V_i shows that the ith sample point exerts more of an effect on the result. In Figure 8.3, samples 2, 4, 28, 30, 33, and 45 are leverage samples.

Practical Advice. If one has a posterior parameter set generated by an MCMC method, then it is easy to numerically calculate ISCV, AIC, DIC, and WAIC. Hence the author recommends that all of them are calculated.

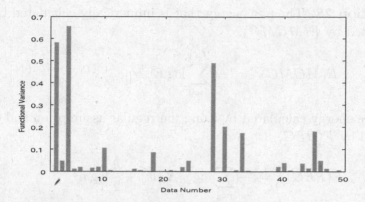

Figure 8.3: Functional variance for each sample point. The horizontal line shows the number of sample point in Figure 8.2. The vertical line shows the partial functional variance V_i of each sample point. The leverage sample points have large partial functional variances.

(1) If they are all equal, then they can be employed.

(2) If a statistical model has a hierarchical structure such as a normal mixture and a neural network, and if ISCV = WAIC >> DIC, then the posterior distribution is not localized. In this case, DIC is not appropriate.

(3) If ISCV $\neq$ WAIC, then there may exist a leverage sample point. A data analyst had better reconsider whether such leverage sample point should be included or not. In conditional independent problems such as time series analysis, ISCV does not correspond to the generalization loss whereas WAIC does.

8.1.3 Criteria for Free Energy

In this subsection, we study the model selection problem by the free energy or the minus log marginal likelihood. If we know the true distribution $q(x)$ and the real log canonical threshold and its multiplicity (λ, m), then the asymptotic free energy is given by

$$F_n = nL_n(w_0) + \lambda \log n - (m-1) \log \log n + O_p(1),$$

where w_0 is the optimal prameter. However, w_0, λ, and m depend on the true distribution $q(x)$, this asymptotic expansion cannot be used directly for estimating F_n.

Definition 28. The free energy that is numerically calculated by eq.(7.19) is denoted by $F(MCMC)$.

$$F(MCMC) \;\; = \;\; -\sum_{j=0}^{J-1} \log E_w^{(\beta_k)}\left[e^{-(\beta_{k+1}-\beta_k)nL_n(w)}\right].$$

The free energy calculated by using the regular assumption and eq.(4.57) is denoted by $F(REG)$,

$$F(REG) \;\; = \;\; -\sum_{i=1}^{n} \log p(X_i|\hat{w}) + \frac{d}{2}\log n$$

$$+\frac{1}{2}\log\det J(\hat{w}) - \log\varphi(\hat{w}) - \frac{1}{2}\log(2\pi), \qquad (8.42)$$

where $\hat{w}$ is the maximum likelihood estimator. The Schwarz BIC is obtained by removing the constant order term from $F(REG)$,

$$\text{BIC} = -\sum_{i=1}^{n}\log p(X_i|\hat{w}) + \frac{d}{2}\log n.$$

The widely applicable Bayesian information criterion WBIC is

$$\text{WBIC} = -\mathbb{E}_w^{(1/\log n)}\left[\sum_{i=1}^{n}\log p(X_i|w)\right],$$

where $\mathbb{E}_w^{(1/\log n)}[\;\;]$ is the posterior average using $\beta = 1/\log n$,

$$\mathbb{E}_w^{(\beta)}[f(w)] = \frac{\displaystyle\int \prod_{i=1}^{n} f(w)\,p(X_i|w)^{\beta}\varphi(w)dw}{\displaystyle\int \prod_{i=1}^{n} p(X_i|w)^{\beta}\varphi(w)dw}.$$

Comparison of $F(MCMC)$, $F(REG)$, BIC, and WBIC. If a true model is regular for a statistical model, then all criteria can be employed. In such a case,

$$\begin{aligned}
F(MCMC) &= F_n + \varepsilon, \\
F(REG) &= F_n + o_p(1), \\
\text{BIC} &= F_n + O_p(1), \\
\text{WBIC} &= F_n + O_p(1), \\
\text{WBIC} &= \text{BIC} + o_p(1),
\end{aligned}$$

where ε depends on numerical calculation. If a true model is not regular for a statistical model, then

$$
\begin{aligned}
F(MCMC) &= F_n + \varepsilon, \\
\text{WBIC} &= F_n + O_p((\log n)^{1/2}),
\end{aligned}
\tag{8.43}
$$

where the second equation is proved in [85]. The mathematical structure is explained in Theorem 25. Note that in Thereom 25, the function $H(w)$ has no random fluctuation. In the case when $H(w)$ is a stochastic process, then the refined proof shows eq.(8.43).

Theorem 25. *Assume that $H(w)$ is an analystic function of w and $\varphi(w)$ is a C^∞ class function. Let F_1 and F_2 be*

$$
F_1 = -\log \int \exp(-nH(w))\varphi(w)dw,
$$

$$
F_2 = \frac{\displaystyle\int nH(w)\,\exp(-(n/\log n)H(w))\varphi(w)dw}{\displaystyle\int \exp(-(n/\log n)H(w))\varphi(w)dw}.
$$

Then, even if the Hessian matrix $\nabla^2 H(w_0)$ at a minimum point w_0 is singular,

$$
F_1 - F_2 = o(\log n).
$$

Proof. If the minimum value of $H(w)$ is H_0, and $H_1(w) = H(w) - H_0$,

$$
F_1 = nH_0 - \log \int \exp(-nH_1(w))\varphi(w)dw,
$$

$$
F_2 = nH_0 + \frac{\displaystyle\int nH_1(w)\,\exp(-(n/\log n)H_1(w))\varphi(w)dw}{\displaystyle\int \exp(-(n/\log n)H_1(w))\varphi(w)dw}.
$$

Hence we can assume $H_0 = 0$ without loss of generality. The zeta function of $H(w)$ is defined by

$$
\zeta(z) = \int H(w)^z \varphi(w)dw \quad (z \in \mathbb{C}).
$$

Then we can derive
(1) In the region $\text{Re}(z) > 0$, $\zeta(z)$ is an analytic function of a complex variable z.

(2) $\zeta(z)$ can be analytically continued to a unique meromorphic function whose poles $(-\lambda_1) > (-\lambda_2) >, ...,$ are all real and negative values. We define m_k as the order of the pole $(-\lambda_k)$. Then $\zeta(z)$ has the Laurent expansion as

$$\zeta(z) = \sum_{k=1}^{\infty} \sum_{m=1}^{m_k} \frac{C_{km}}{(z + \lambda_k)^m},$$

where $C_{km} \in \mathbb{C}$. The state density and partition functions are respectively defined by

$$v(t) = \int \delta(t - f(x))\varphi(x)dx \quad (0 < t < 1),$$

$$Z(n) = \int \exp(-nf(x))\varphi(x)dx \quad (n > 0).$$

Then it follows that

$$\zeta(z) = \int_0^1 t^z \, v(t) \, dt,$$

$$Z(n) = \int_0^1 \exp(-nt) \, v(t) \, dt.$$

In other words, $\zeta(z)$ and $Z(n)$ are the Mellin and Laplace transforms of $v(t)$, respectively. The following equation can be derived by mathematical induction about $m = 1, 2, ...$

$$\frac{1}{(z + \lambda)^m} = \frac{1}{(m - 1)!} \int_0^1 t^{\lambda-1}(\log t)^{m-1} \, t^z \, dt.$$

By using this equation and the Laurent expansion of $\zeta(z)$, we obtain the asymptotic expansion,

$$v(t) = \sum_{k=1}^{\infty} \sum_{m=1}^{m_k} \frac{C_{km}}{(m - 1)!} t^{\lambda_k-1} (\log t)^{m-1}.$$

Therefore, the asymptotic expansion of $Z(n)$ holds,

$$Z(n) = \sum_{k=1}^{\infty} \sum_{m=1}^{m_k} \frac{C_{km}}{(m - 1)!} \int_0^1 t^{\lambda_k-1}(\log t)^{m-1} \, \exp(-nt) \, dt$$

$$= \sum_{k=1}^{\infty} \sum_{m=1}^{m_k} \frac{C_{km}}{(m - 1)!} \int_0^n (t/n)^{\lambda_k-1}(\log(t/n))^{m-1} \, \exp(-t) \, \frac{dt}{n}.$$

In the case $n \to \infty$, the largest order term is

$$Z(n) \cong \frac{C_{11}\Gamma(\lambda_1)}{(m-1)!} \cdot \frac{(\log n)^{m_1-1}}{n^{\lambda_1}}.$$

Hence

$$F_1 = \lambda_1 \log n - (m_1 - 1)\log\log n + \text{Const.} + \cdots$$

On the other hand, by using $\beta = n/\log n$, we define

$$F_{21} = \int_0^1 \exp(-\beta t)\, nt\, v(t)\, dt,$$

$$F_{22} = \int_0^1 \exp(-\beta t)\, v(t)\, dt.$$

Then $F_2 = F_{21}/F_{22}$. By using the same method as F_1,

$$F_{21} \cong \frac{C_{11}\Gamma(\lambda_1+1)}{(m-1)!} \cdot \frac{n(\log\beta)^{m_1-1}}{\beta^{\lambda_1+1}},$$

$$F_{22} \cong \frac{C_{11}\Gamma(\lambda_1)}{(m-1)!} \cdot \frac{(\log\beta)^{m_1-1}}{\beta^{\lambda_1}}.$$

Then by $\Gamma(\lambda_1 + 1) = \lambda_1\Gamma(\lambda_1)$, it follows that

$$F_2 \cong \lambda_1 \log n,$$

which completes the theorem. □

Example 57. A simple model selection experiment using WBIC was conducted. Let $x \in \mathbb{R}^2$, $y \in \mathbb{R}$. The input X_i was generated from the uniform distribution of $[-2,2]^2$. The true distribution of Y_i was set as $p(y|x, w_0)$ where $p(y|x, w_0)$ was made by a neural network defined by eq.(2.27) with three hidden units $H = 3$. From this true distribution, $n = 500$ sample points were generated. A prior was set by the normal distribution $\mathcal{N}(0, 10^2)$ for each u_{jk} and $w_{k\ell}$. The 1000 posterior parameters were approximated by a Metropolis method with the burn-in 1000 and sampling interval 200. Figure 8.4 shows WBICs for a neural network with $H = 1, 2, 3, 4, 5$. As in a figure, a true model could be chosen by WBIC.

Remark 67. In general, the asymptotic form of the free energy or the minus log marginal likelihood depends on a true distribution, since the real log canonical threshold depends on the true distribution. Recently, a new method was devised by which both the true distribution and the free energy can estimated simultaneously using the real log canonical threshold [19].

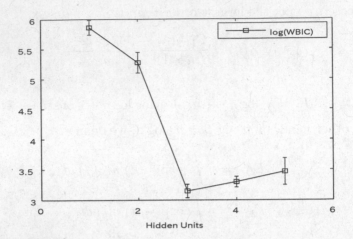

Figure 8.4: WBIC of neural networks. The horizontal line shows the number of hidden units in a neural network, and the vertical line WBIC. By using WBIC, a model selection according to the asymptotic free energy can be realized.

8.1.4   Discussion for Model Selection

Let us study model selection problems from two different points of view. This discussion is based on Professor Akaike's argument.

Artificial case. Assume that a true distribution is realizable by a statistical model which is contained in the finite set of candidate models. Since such a case is rare in the real world, it is called an artificial case. In the artificial case, the minimal model by which a true distribution is realizable is called the true model. A model selection algorithm is called *consistent*, if the probability that the selected model is equal to the true model converges to one for $n \to \infty$. In general, model selection algorithms which employ the cross validation loss, AIC, DIC, and WAIC are not consistent. The reason why they are inconsistent is that random fluctuation according to a sample is in proportion to the difference of the generalization loss. On the other hand, the model selection algorithm which is based on the free energy is consistent, because the main order part $\log n$ is not a random variable but a constant which is larger than the random fluctuation. Therefore, in an artificial case, the free energy is better than the generalization loss.

Natural case. Assume that we have candidate models, but, a true distribution is not realizable by any statistical model whose parameter has finite dimension. Almost all statistical problems in the real world are classified into this case, hence are called the natural cases. In a natural case, if the number of random variables increases, then the best model also becomes more complex. Consistency has no meaning. A model selection algorithm is called *efficient*, if the average generalization loss of the selected model is minimized among the candidate models. From the view point of efficiency, the generalization loss is better than the free energy.

When we compare several model selection problems by computer simulation, we often set a true distribution as an artificial case. However, such an experiment may be different from the natural cases.

8.2 Hyperparameter Optimization

A parameter of a prior distribution is called a hyperparameter. In this section, we study several problems in hyperparameter optimization.

Remark 68. If a set of a statistical model $p(x|w)$ and a prior $\varphi(w|\theta)$ is prepared, one might think the hyperparameter θ could be automatically optimized by intoducing the hyperprior distribution $\varphi_1(\theta)$. However, it is not true. If the hyperprior distribution is employed, then it strongly affects the optimal hyperparameter, resulting that the chosen hyperparameter is not optimized but detemined by the choice of the hyperprior. In such a case a prior $\int \varphi(w|\theta)\varphi_1(\theta)d\theta$ should be evaluated as a prior. Hence we need a method how to evaluate $(p(x|w), \varphi(w))$. For example, in a nonparametric Bayesian estimation, the Dirichlet hyperparameter α might be determined by using the hyperprior, but it is not the automatically optimal one. Even in nonparametric cases, the cross validation and WAIC can be employed to evaluate the hyperparameter.

In this section, we study the general case

$$\varphi(w) \geq 0, \quad \text{but } \int \varphi(w)dw \quad \text{may be infinite.}$$

Even in such cases, we can use the same definition of the posterior distribution as the case $\int \varphi(w)dw = 1$,

$$p(w|X^n) = \frac{\varphi(w)\displaystyle\prod_{i=1}^{n} p(X_i|w)}{\displaystyle\int \varphi(w)\prod_{i=1}^{n} p(X_i|w)dw},$$

because this definition does not require any normalizing condition of $\varphi(w)$. Hence the definitions of the generalization, cross validation, and training losses are also invariant. However, the free energy should be redefined by

$$F_n = -\log \int \prod_{i=1}^{n} p(X_i|w)\varphi(w)dw + \log \int \varphi(w)dw,$$

because, without the second term in the right hand side, F_n is an unbounded function of φ.

Remark 69. (1) In the hyperparameter optimization problem about the generalization loss, we admit cases when $\int \varphi(w)dw = \infty$. For example, $\varphi(w) = 1$ on the unbounded parameter set W can be used in this section. On the other hand, for the free energy, $\int \varphi(w)dw < \infty$ is necessary for finite F_n. That is to say, the generalization loss and the free energy have the essential difference in preparing the set of priors.

(2) Let $\hat{w}$ be the maximum likelihood estimator. Then, for an arbitrary w

$$\prod_{i=1}^{n} p(X_i|w) \le \prod_{i=1}^{n} p(X_i|\hat{w}).$$

Hence if there exists a sequence of priors $\{\varphi_k(w)\}$ such that $\varphi_k(w) \to \delta(w - \hat{w})$, then the infimum value of F_n is attained by such a sequence, which converges to the maximum likelihood method. Therefore, when the free energy is applied to the prior optimization, the set of candidate priors should be set so as that such a sequence is not contained.

Example 58. The Dirichlet distribution

$$\varphi(a) \propto a^{\alpha-1}(1-a)^{\beta-1},$$

converges to $\delta(a - a_0)$ by

$$\begin{aligned} \alpha_k &= ka_0 \\ \beta_k &= k(1-a_0) \end{aligned}$$

and $k \to \infty$. Hence α and β should be bounded by some constant.

8.2.1 Criteria for Generalization Loss

In the model selection problem according to the generalization loss, we studied the cross validation, AIC, TIC, DIC, and WAIC. Since the effect of the prior choice to the generalization loss is weaker than that of the statistical model, we need a precise tool to observe the difference of the small order. In fact, neither AIC, TIC, nor DIC can be applied to prior evaluation. In this subsection, we study the hyperparameter optimization by the cross validation and WAIC. If a true distribution is regular for a statistical model, then we have the following theorem even if a true distribution is not realizable by a statistical model.

Regular case. In regular cases, the effect of the hyperparameter optimization by the cross validation and WAIC are mathematically clarified. Let $\varphi_0(w)$ and $\varphi(w)$ be arbitrary fixed and candidate priors respectively. As a typical case, $\varphi_0(w) \equiv 1$ for all $w \in W$ can be chosen. The empirical log loss function and the maximum *a posteriori* (MAP) estimator $\hat{w}$ using $\varphi_0(w)$ are respectively defined by

$$L_n(w) = -\frac{1}{n} \sum_{i=1}^{n} \log p(X_i|w) - \frac{1}{n} \log \varphi_0(w), \qquad (8.44)$$

$$\hat{w} = \arg\min_{w \in W} L_n(w), \qquad (8.45)$$

where either $L_n(w)$ or $\hat{w}$ does not depend on the candidate prior $\varphi(w)$. If $\varphi_0(w) \equiv 1$, then $\hat{w}$ is equal to the maximum likelihood estimator (MLE). The average log loss function and the parameter that minimizes it are respectively defined by

$$L(w) = -\int q(x) \log p(x|w) dx, \qquad (8.46)$$

$$w_0 = \arg\min_{w \in W} L(w). \qquad (8.47)$$

In this section, we use the following notations for simple description.
(1) A parameter is denoted by $w = (w^1, w^2, ..., w^k, ..., w^d) \in \mathbb{R}^d$.
(2) For an arbitrary function $f(w)$ and nonnegative integers $k_1, k_2, ..., k_m$, we define

$$(f)_{k_1 k_2 \cdots k_m}(w) = \frac{\partial^m(f)}{\partial w^{k_1} \partial w^{k_2} \cdots \partial w^{k_m}}(w). \qquad (8.48)$$

(3) We adopt Einstein's summation convention and $k_1, k_2, k_3, ...$ are used for

such suffixes. For example,

$$X_{k_1 k_2} Y^{k_2 k_3} = \sum_{k_2=1}^{d} X_{k_1 k_2} Y^{k_2 k_3}.$$

In other words, if a suffix k_i appears upper and lower, it means automatic summation over $k_i = 1, 2, ..., d$. In this section, for each k_1, k_2, $X^{k_1 k_2} = X_{k_2}^{k_1} = X_{k_1 k_2}$.

Definition. (Empirical mathematical relations between priors) For a fixed and candidate priors $\varphi_0(w)$ and $\varphi(w)$, the prior ratio function is defined by

$$\phi(w) = \varphi(w)/\varphi_0(w).$$

The empirical mathematical relation between two priors at a parameter w is defined by

$$
\begin{aligned}
M(\phi, w) \;=\;& A^{k_1 k_2} (\log \phi)_{k_1} (\log \phi)_{k_2} + B^{k_1 k_2} (\log \phi)_{k_1 k_2} \\
& + C^{k_1} (\log \phi)_{k_1},
\end{aligned} \tag{8.49}
$$

where

$$J^{k_1 k_2}(w) \;=\; \text{Inverse matrix of } (L_n)_{k_1 k_2}(w), \tag{8.50}$$

$$A^{k_1 k_2}(w) \;=\; \frac{1}{2} J^{k_1 k_2}(w), \tag{8.51}$$

$$B^{k_1 k_2}(w) \;=\; \frac{1}{2}(J^{k_1 k_2}(w) + J^{k_1 k_3}(w) J^{k_2 k_4}(w) F_{k_3, k_4}(w)), \tag{8.52}$$

$$
\begin{aligned}
C^{k_1}(w) \;=\;& J^{k_1 k_2}(w) J^{k_3 k_4}(w) F_{k_2 k_4, k_3}(w) \\
& -\frac{1}{2} J^{k_1 k_2}(w) J^{k_3 k_4}(w)(L_n)_{k_2 k_3 k_4}(w) \\
& -\frac{1}{2} J^{k_1 k_2}(w) J^{k_3 k_4}(w) J^{k_5 k_6}(w) \\
& \times (L_n)_{k_2 k_3 k_5}(w) F_{k_4, k_6}(w),
\end{aligned} \tag{8.53}
$$

and

$$F_{k_1, k_2}(w) \;=\; \frac{1}{n} \sum_{i=1}^{n} (\log p(X_i|w))_{k_1} (\log p(X_i|w))_{k_2}, \tag{8.54}$$

$$F_{k_1 k_2, k_3}(w) \;=\; \frac{1}{n} \sum_{i=1}^{n} (\log p(X_i|w))_{k_1 k_2} (\log p(X_i|w))_{k_3}. \tag{8.55}$$

Remark. Note that neither $A^{k_1 k_2}(w)$, $B^{k_1 k_2}(w)$, nor $C^{k_1}(w)$ depends on a candidate prior $\varphi(w)$. Therefore $M(\phi, w)$ is determined by only $log\phi$ as a function of the candidate prior.

Definition. (Average mathematical relations of priors) The average mathematical relation $\mathcal{M}(\phi, w)$ is defined by the same manner as eq.(8.49) by replacement

$$J^{k_1 k_2}(w) \;\;\mapsto\;\; \text{Inverse matrix of } \mathbb{E}[(L_n)_{k_1 k_2}(w)], \qquad (8.56)$$

$$(L_n)_{k_1 k_2}(w) \;\;\mapsto\;\; \mathbb{E}[(L_n)_{k_1 k_2}(w)], \qquad (8.57)$$

$$(L_n)_{k_1 k_2 k_3}(w) \;\;\mapsto\;\; \mathbb{E}[(L_n)_{k_1 k_2 k_3}(w)], \qquad (8.58)$$

$$F_{k_1, k_3}(w) \;\;\mapsto\;\; \mathbb{E}[F_{k_1 k_2}(w)], \qquad (8.59)$$

$$F_{k_1 k_2, k_3}(w) \;\;\mapsto\;\; \mathbb{E}[F_{k_1 k_2, k_3}(w)]. \qquad (8.60)$$

The following theorem shows the effect of the choice of the candidate prior by comparison of the fixed prior.

Theorem 26. *Let $\varphi_0(w)$ and $\varphi(w)$ be fixed and candidate priors respectively. The prior ratio function is defined by*

$$\phi(w) = \varphi(w)/\varphi_0(w).$$

Let $M(\phi, w)$ and $\mathcal{M}(\phi, w)$ be the empirical and average mathematical relations between $\varphi(w)$ and $\varphi_0(w)$. As random variables,

$$\mathrm{CV}(\varphi) \;=\; \mathrm{CV}(\varphi_0) + \frac{M(\phi, \hat{w})}{n^2} + O_p(\frac{1}{n^3}), \qquad (8.61)$$

$$\mathrm{WAIC}(\varphi) \;=\; \mathrm{WAIC}(\varphi_0) + \frac{M(\phi, \hat{w})}{n^2} + O_p(\frac{1}{n^3}), \qquad (8.62)$$

$$\mathrm{CV}(\varphi) \;=\; \mathrm{WAIC}(\varphi) + O_p(\frac{1}{n^3}). \qquad (8.63)$$

Their expected values satisfy

$$\mathbb{E}[\mathrm{CV}(\varphi)] \;=\; \mathbb{E}[\mathrm{CV}(\varphi_0)] + \frac{\mathcal{M}(\phi, w_0)}{n^2} + O(\frac{1}{n^3}), \qquad (8.64)$$

$$\mathbb{E}[\mathrm{WAIC}(\varphi)] \;=\; \mathbb{E}[\mathrm{WAIC}(\varphi_0)] + \frac{\mathcal{M}(\phi, w_0)}{n^2} + O(\frac{1}{n^3}), \qquad (8.65)$$

where

$$M(\phi, \hat{w}) \;=\; \mathcal{M}(\phi, w_0) + O_p(\frac{1}{n^{1/2}}), \qquad (8.66)$$

$$\mathbb{E}[M(\phi, \hat{w})] \;=\; \mathcal{M}(\phi, w_0) + O(\frac{1}{n}). \qquad (8.67)$$

On the other hand, the generalization loss satisifies

$$G(\varphi) = G(\varphi_0) + \frac{1}{n}(\hat{w}^{k_1} - (w_0)^{k_1})(\log\phi)_{k_1}(\hat{w}) + O_p(\frac{1}{n^2})$$

$$= G(\varphi_0) + O_p(\frac{1}{n^{3/2}}), \tag{8.68}$$

$$\mathbb{E}[G(\varphi)] = \mathbb{E}[G(\varphi_0)] + \frac{\mathcal{M}(\phi, w_0)}{n^2} + O(\frac{1}{n^3}). \tag{8.69}$$

For the proof of this theorem, see [86]. If a candidate prior has a hyperparameter θ, which is written as $\varphi(w) = \varphi(w|\theta)$, the following facts are derived by this theorem.

(1) The hyperparameters that minimize $\mathbb{E}[CV]$, $\mathbb{E}[WAIC]$, and $\mathbb{E}[G_n]$ are symptotically equal to each other.

(2) The hyperparameters that minimize CV, WAIC, and $\mathbb{E}[G_n]$ are asymptotically equal to each other. Hence by minimizing CV or WAIC, we can find the optimal hyperparameter that minimizes $\mathbb{E}[G_n]$ asymptotically.

(3) [**Important point**]. The hyperparameters that minimize the random variable G_n and the average $\mathbb{E}[G_n]$ are not equal to each other even asymptotically. In general they are far from each other and one does not converge to the other even if n tends to infinity. By minimizing CV or WAIC, we can find the optimal hyperparameter that minimizes $\mathbb{E}[G_n]$, but we cannot find the optimal hyperparameter that minimizes G_n.

Hence by determining the hyperparameter by minimizing the cross validation or WAIC, $\mathbb{E}[G_n]$ is asymptotically minimized but G_n is not. (see Example. 59). It is strongly conjectured that there is no observable which can estimate the random variable G_n for an arbitrary true distribution, because we do not know the true distribution. This is the conjecture about the limit of statistical estimation.

Remark 70. Since $\mathbb{E}[CV(\varphi_0)]$ of X^n is equal to $\mathbb{E}[G(\varphi_0)]$ of X^{n-1} and

$$\frac{1}{n-1} - \frac{1}{n} = \frac{1}{n^2} + o(\frac{1}{n^2}),$$

it immediately follows from Theorem 26 that

$$\mathbb{E}[G(\varphi)] = \mathbb{E}[G(\varphi_0)] + \frac{\mathcal{M}(\phi, w_0)}{n^2} + o(\frac{1}{n^2}), \tag{8.70}$$

$$\mathbb{E}[CV(\varphi)] = \mathbb{E}[G(\varphi_0)] + \frac{d/2 + \mathcal{M}(\phi, w_0)}{n^2} + o(\frac{1}{n^2}), \tag{8.71}$$

$$\mathbb{E}[WAIC(\varphi)] = \mathbb{E}[G(\varphi_0)] + \frac{d/2 + \mathcal{M}(\phi, w_0)}{n^2} + o(\frac{1}{n^2}). \tag{8.72}$$

Nonregular case. In nonregular cases, determination of the hyperparameter is the essential procedure of Bayesian inference. However, mathematical analysis for this case is still difficult, because nonregular statistical models have phase transitions according to the hyperparameter controlling. For estimating the averages, the cross validation and WAIC can be employed. See Example 67.

8.2.2   Criterion for Free Energy

For the purpose of the minimization of the free energy, neither BIC nor WBIC can be applied, because they do not estimate the constant order term. The values $F(REG)$ and $F(MCMC)$ can be used in regular and all cases, respectively. If $F(REG)$ is employed, then the hyperparameter that minimizes the $F(REG)$ is equal to the one that maximizes $\varphi(\hat{w})$. In other words, the hyperparameter is optimized so that the prior at the maximum likelihood estimator is maximized.

For the hyperparameter optimization, there is an another method. Let $F_n(\alpha)$ be the free energy for a prior $\varphi(w|\alpha)$, where α is a hyperparameter. Then

$$\frac{dF_n}{d\alpha} = \frac{\int \left(-\frac{\partial}{\partial \alpha} \log \varphi(w|\alpha)\right) \varphi(w|\alpha) \prod_{i=1}^{n} p(X_i|w)dw}{\int \varphi(w|\alpha) \prod_{i=1}^{n} p(X_i|w)dw}$$

$$= -\mathbb{E}_w\left[\frac{\partial}{\partial \alpha} \log \varphi(w|\alpha)\right].$$

Hence using the increase and decrease table, the hyperparameter that minimizes $F_n(\alpha)$ can be found. In order to calculate $F_n(MCMC)$, we need all posterior distributions for many inverse temperatures, whereas $dF_n/d\alpha$ can be calculated by one MCMC process.

Example 59. By using a statistical model which enables us to exactly calculate the generalization and cross validation losses and the free energy, let us study the hyperparameter optimization problem numerically. We use the normal distribution and its conjugate prior defined by eqs.(2.1) and (2.2). Let $n = 200$. The hyperparameter (ϕ_1, ϕ_2, ϕ_3) in the region

$$0 < \phi_1 < 10$$

is examined, where $\phi_2 = 0$ and $\phi_3 = 1$. We conducted 100 independent experiments. In Figure 8.5, the horizontal line shows the value ϕ_1.

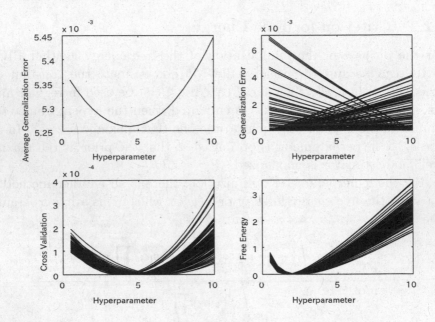

Figure 8.5: Hyperparameter optimization. The horizontal lines in all figures show the value of the hyperparameter. The vertical lines show the average generalization error, the generalization error, the cross validation error, and the free energy. The minimum points of the free energy is not equal to that of the average generalization error. The minimum point of the generalization error has very large variance, and thus it is not equal to that of the average generalization error. The minimum point of the cross validation error is asymptotically equal to that of the average generalization error.

(1) Upper left: The average generalization error for a given hyperparameter.
(2) Upper right: Generalization errors for a given hyperparameter.
(3) Lower left: Cross validation losses for a given hyperparameter.
(4) Lower right: Free energies for a given hyperparameter.
In (2), (3), and (4), each function of a hyperparameter is displayed by calibration that the minimum value is equal to zero. The hyperparameter that minimizes the average generalization loss is almost equal to $\phi_1 = 5$. Each hyperparameter that minimizes each generalization loss strongly depends on a sample X^n. It almost always lies on the outside of $0 < \phi_1 < 10$. Note that it sometimes lies in $\phi_1 < 0$. The hyperparameter that minimizes the cross validation loss is in the neighborhood of $\phi_1 = 5$. The hyperparameter that minimizes the free energy is in the neighborhood of $\phi_1 = 2$. These results show the case $n = 200$. If n is smaller, the variance of the chosen hyperparameter is larger, hence too much optimization of the hyperparameter may be dangerous. Note that if all of (ϕ_1, ϕ_2, ϕ_3) are optimized simultaneously, then the hyperparameter diverges.

8.2.3 Discussion for Hyperparameter Optimization

Regular case. Assume a true distribution is regular for a statistical model, then:
(1) If a hyperparameter is optimized by minimization of the cross validation or WAIC, then the average generalization loss is minimized asymptotically. However, the generalization loss itself is not minimized. Moreover the random fluctuation of the optimized hyperparameter is not small.
(2) If a hyperparameter is optimized by minimizing the free energy, then it is asymptotically equivalent to maximizing the value of the prior at the maximum likelihood estimator. The random fluctuation of the optimized hyperparameter may be smaller than the cross validation, however, it does not minimize the generalization loss.
Therefore, even if the prior is optimized, its effect to the accurate prediction is small. Moreover, the random fluctuation may make the variance of the optimized parameter larger. Therefore, too much optimization is not necessary. However, choosing the appropriate prior among several candidates by the cross validation or WAIC may be useful.
Singular case. Assume that a true distribution is singular for a statistical model. Then both the generalization loss and the free energy have the phase transitions for hyperparameter controlling (for the definition of the phase transition, see the following chapter). If the real log canonical threshold is minimized by appropriate hyperparameter choosing, then it makes

the generalization loss smaller. However, in singular cases, the effect of control hyperparameter to the precise prediction is not sufficiently clarified mathematically. This is an important problem for the future study.

Example 60. (LASSO) Let us study LASSO (least absolute shrinkage and selection operator). Let $x \in \mathbb{R}^M$, $y \in \mathbb{R}^N$. A model and a prior are defined by

$$p(y|x, w) \;\propto\; \exp(-\frac{1}{2\sigma^2}\|y - wx\|^2),$$

$$\varphi(w|\ell) \;\propto\; \exp(-\ell \sum_{jk} |w_{jk}|),$$

where $w = \{w_{jk}\}$ is a $N \times M$ matrix, σ is a constant, and ℓ is a hyperparameter. The purpose of using this prior is to make the estimated parameter sparse. In fact, the maximum *a posteriori* (MAP) estimator by using this prior becomes sparse by choosing ℓ appropriately. Let us study the Bayesian case. Assume that the true distribution is $q(y|x, w_0)$. Then the Bayesian generalization error is

$$\mathbb{E}[G_n] = nS + \frac{\lambda}{n} + o(1/n),$$

where S is the entropy of $p(y|x, w_0)$ and λ is the real log canonical threshold. The value $(-\lambda)$ is equal to the maximum pole of the zeta function,

$$\zeta(z) = \int \|w - w_0\|^{2z} \exp(-\ell \sum_{jk} |w_{jk}|)dw.$$

Even if almost all elements of w_0 are equal to zero, the largest pole of the zeta function is equal to $-d/2$, where $d = MN$ is the dimension of the parameter, since $\varphi(w_0|\ell) > 0$. In other words, $\lambda = d/2$ does not depend on the choice of ℓ. Therefore, in Bayesian estimation, the prior $\exp(-\ell \sum_{jk} |w_{jk}|)$ is not appropriate for sparse representation of the parameter. In LASSO, Bayesian estimation is very different from MAP estimation.

Example 61. (Bayesian LASSO) Let $x \in \mathbb{R}^M$, $y \in \mathbb{R}$, $w \in \mathbb{R}^M$. We study a statistical model

$$p(y|x, w) = \frac{1}{(2\pi\sigma^2)^{1/2}} \exp\Big(-\frac{1}{2\sigma^2}(y - w \cdot x)^2\Big),$$

where $\sigma > 0$ is not a parameter but a constant, and a prior

$$\varphi(w) = C \prod_{j=1}^M \frac{1}{|w_j|^a} \exp(-\varepsilon w_j^2),$$

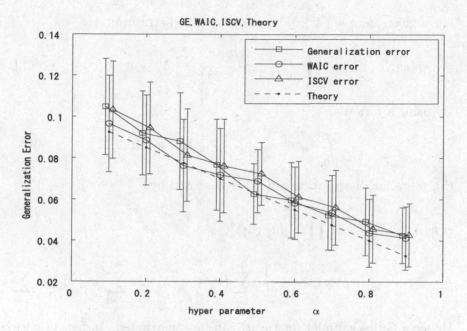

Figure 8.6: Generalization error by Bayesian LASSO. The horizontal line shows the hyperparameter α. The generalization error, the cross validation error, and WAIC error are compared with the theoretical value. By using Bayesian LASSO, the generalization error can be made smaller, if the true parameter is sparse.

where $C(a, \epsilon)$ is a constant

$$C(a, \epsilon) = \left(\frac{\varepsilon^{(1-a)/2}}{\Gamma((1-a)/2)} \right)^M,$$

for a given hyperparameter $a < 1$ and $\varepsilon > 0$. Note that as a becomes large, the posterior distribution concetrates on the neighborhood of the origin. Assume that the true distribution is $p(y|x, w_0)$, where the number of the nonzero elements of w_0 is equal to M_0. The true distribution $q(x)$ is the direct product of the standard normal distribution. Then the real log canonical threshold is

$$\lambda(a) = \frac{1}{2}\{M_0 + (1 - a)(M - M_0)\}, \tag{8.73}$$

resulting that the asymptotic generalization error is

$$\mathbb{E}[G_n] - S = \frac{\lambda(a)}{n} + o(1/n).$$

For a given sample (X^n, Y^n), the posterior distribution is

$$p(w|X^n, Y^n) \propto \prod_{j=1}^{M} \frac{1}{|w_j|^a} \exp\left(-\frac{1}{2\sigma^2} \sum_{i=1}^{n} (Y_i - w \cdot X_i)^2 - \varepsilon \sum_{j=1}^{M} w_j^2\right).$$

By using a formula,

$$\int_0^\infty u^{a/2-1} \exp(-w^2 u) du = \frac{\Gamma(a/2)}{|w|^a},$$

the posterior distribution can be represented by

$$p(w|X^n, Y^n) \propto \left(\prod_{j=1}^{M} \int du_j \, (u_j)^{a/2-1}\right)$$

$$\times \exp\left(-\frac{1}{2\sigma^2} \sum_{i=1}^{n} (Y_i - w \cdot X_i)^2 - \sum_{j=1}^{M} (w_j)^2 u_j - \varepsilon \sum_{j=1}^{M} w_j^2\right).$$

Hence a Gibbs sampler for (w, u) can be constructed. In fact $p(w|u)$ is the normal distribution whose average is

$$m = S^{-1}\left(\frac{1}{\sigma^2} \sum_{i=1}^{n} Y_i X_i\right)$$

and covariance matrix is S^{-1}. Here

$$S = \frac{1}{\sigma^2} \sum_{i=1}^{n} X_i (X_i)^T + 2 Diag(\varepsilon + u_1, \varepsilon + u_2, ..., \varepsilon + u_M),$$

where $Diag(u_1, u_2, ..., u_M)$ is the diagonal matrix whose diagonal coefficients are $(u_1, u_2, ..., u_M)$. On the other hand, $p(u|w)$ is the direct product of the gamma distribution $\mathcal{G}(u_j|a/2, 1/(w_j)^2)$, where

$$\mathcal{G}(x|a, b) = \frac{1}{b^a \Gamma(a)} x^{a-1} \exp(-\frac{x}{b}). \tag{8.74}$$

Figure 8.6 shows an experimental result for the case $M = 40$, $M_0 = 10$, and $n = 200$. In order to make the posterior distribution stable, if $|u_j| \geq u_{max} = 1000000$, then it is replaced by $\pm u_{max}$. The horizontal line shows the hyperparamater α and the generalization error, the cross validaiton error, and WAIC error are compared with the theoretical value. In this case the true parameter is sparse, hence the generalization error is made smaller by using Bayesian LASSO.

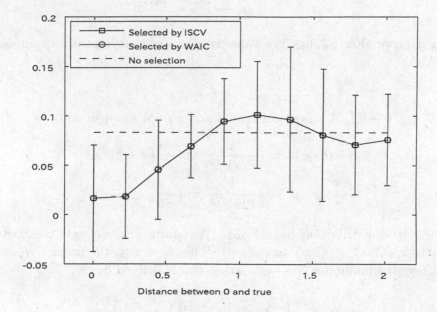

Figure 8.7: Generalization errors by the minimum cross validation. The solid line shows the generalization error by the model selection with respect to the minimum cross validation loss. The dotted line shows that of the unselected larger model. In this experiment, WAIC and CV resulted in the same model selection. Note that the model selection does not always minimize the generalization error. In fact, in the delicate case when two models are almost balanced, the generalization error becomes larger.

8.3   Problems

1. Let U_n be a random variable which is defined by

$$U_n = \frac{1}{n} \sum_{i=1}^{n} \log \mathbb{E}_w[p(X_i|w)]$$

$$- \frac{2}{n} \sum_{i=1}^{n} \mathbb{E}_w[\log p(X_i|w)].$$

Then prove that U_n has the same second order asymptotic expansion as WAIC.

2. Let $x, a \in \mathbb{R}^d$. A statistical model and a prior are defined by

$$p(x|a) = \frac{1}{(2\pi)^{d/2}} \exp(-\frac{1}{2}\|x - a\|^2),$$

$$\varphi(a) = \frac{1}{(2\pi)^{d/2}} \exp(-\frac{1}{2}\|a\|^2).$$

Let a true distribution be $p(x|a_0)$. We study a model selection between statistical models $p(x|a)$ and $p(x|0)$. That is to say, the predictive density $\hat{p}(x)$ by the minimum cross validation loss is defined by

$$\hat{p}(x) = \begin{cases} \mathbb{E}_w[p(x|w)] & (\text{if } C_n^{(1)} \le C_n^{(0)}) \\ p(x|0) & (\text{otherwise} \end{cases}, \tag{8.75}$$

where $C_n^{(1)}$ is the cross validation loss of $p(x|a)$ and

$$C_n^{(0)} = -\frac{1}{n} \sum_{i=1}^{n} \log p(X_i|0).$$

The generalization error of the minimum cross validation loss is defined by

$$\int p(x|a_0) \log \frac{p(x|a_0)}{\hat{p}(x)} dx.$$

Then this is a function of the distance between 0 and a_0. The solid line in Figure 8.7 shows its behavior as such a function in the case $d = 5$. The horizontal line shows the distance between the origin and the true parameter. The dotted line shows the generalization error of the predicitive density of

$p(x|a)$ without model selection. Discuss the effect of the model selection on the generalization error.

3. If the posterior distribution can be approximated by some normal distribution, then the difference between BIC and the free energy is a constant order term. If otherwise, the difference between WBIC and the free energy is at most a $\log \log n$ order term. Discuss how much such diferences affect the selected models.

4. Assume that the posterior distribution can be approximated by some normal distribution. Then the hyperparameter that minimizes the generalization loss does not converge to the hyperparameter that minimizes the average generalization loss. On the other hand, the hyperparameter that minimizes the cross validation loss or WAIC converges to one which minimizes the average generalization loss. Discuss the best procedure that a statistician can follow to find the minimum generalization error.

5. A neural network which has a deep hierarchical structure has many parameters and the posterior distribution can seldom be approximated by any normal distribution. For such statistical models, Bayesian estimation makes the generalization loss very small. However, it is difficult to approximate the posterior distribution by MCMC. Discuss the best procedure that a statistician can follow for deep learning.

6. Prove eq.(8.73).

Chapter 9

Topics in Bayesian Statistics

In this chapter we research mathematical bases of several topics in Bayesian statistics.

(1) The formal optimality of the Bayesian estimation is explained.

(2) A method how to construct the Bayesian hypothesis test is explained.

(3) The Bayesian model comparison method which is different from the Bayesian hypothesis test is examined,

(4) The concept of the phase transition of the posterior distribution is introduced.

(5) In a statatistical model which has singularities in the parameter space, if the sample size n is small, the posterior distribution is singular, whereas, if n becomes large, it becomes regular. This phenomenon is a kind of phase transition called the discovery process.

(6) In hierarchical Bayesian estimation, we find several different kinds of predictions. There are different cross validation losses and information criteria according to the different predictive losses.

9.1   Formal Optimality

If we know the true prior and the true model, then Bayesian inference is optimal. In this section we confirm this fact.

Assume that $\Phi(w)$ is the true prior density of the parameter w and that $P(x|w)$ is the true conditional density of x. In this book, we assume that the true distribution is unknown, however, in this section, we study the special case that a random parameter W is generated from $\Phi(w)$, then a sample

$X^n = (X_1, X_2, ..., X_n)$ is independently generated from $P(x|w)$.

$$W \ \sim \ \Phi(w),$$
$$X_1, X_2, ..., X_n \ \sim \ P(x|w).$$

The simultaneous probability density function of (W, X^n) is equal to

$$P(w, x^n) = \Phi(w) \prod_{i=1}^{n} P(x_i|w).$$

Therefore a conditional probability density of w for a given sample x^n is equal to

$$P(w|x^n) = \frac{\Phi(w) \prod_{i=1}^{n} P(x_i|w)}{\int dw' \ \Phi(w') \prod_{i=1}^{n} P(x_i|w')}. \tag{9.1}$$

This is equal to the Bayesian posterior distribution for the case when $\Phi(w)$ and $P(x|w)$ are chosen as a prior and a statistical model. The probability distribution of a new x is given by

$$P(x|x^n) = \frac{\int dw \ \Phi(w)P(x|w) \prod_{i=1}^{n} P(x_i|w)}{\int dw' \ \Phi(w') \prod_{i=1}^{n} P(x_i|w')}. \tag{9.2}$$

Also this is equal to the Bayesian predictive distribution for the case when $\Phi(w)$ and $P(x|w)$ are chosen as a prior and a statistical model. Let us prove that this prediction minimizes the average Kullback-Leibler distance under the circumstance $P(w, x^n)$.

Let $f(x|X^n)$ be an arbitrary conditional density function of x for X^n. For a given (w, x^n), the Kullback-Leibler distance from $P(x|w)$ to $f(x|x^n)$ is equal to

$$G(f|w, x^n) = \int dx P(x|w) \log \frac{P(x|w)}{f(x|x^n)}.$$

Let us define an average functional loss of f be

$$\mathcal{G}(f) \ = \ \int dw \ \Phi(w) \left[\prod_{i=1}^{n} \int P(x_i|w)dx_i \right] G(f|w, x^n).$$

Then the function f that minimizes $\mathcal{G}(f)$ gives the optimal inference under the circumstance $P(w, x^n)$. The following theorem shows that such a function is the Bayesian predictive distribution.

Theorem 27. *The average functional loss $\mathcal{G}(f)$ is minimized if and only if $f(x|X^n) = P(x|X^n)$.*

Proof. The function $\mathcal{G}(f)$ is minimized if and only if

$$
\begin{aligned}
\mathcal{G}_0(f) &= -\int dw \Phi(w) \prod_{i=1}^{n} \int P(x_i|w) dx_i \int dx P(x|w) \log f(x|x^n) \\
&= -\prod_{i=1}^{n} \int dx_i \int dx P(x, x^n) \log f(x|x^n)
\end{aligned}
$$

is minimized, where $P(x, x^n)$ is the simultaneous density of (x, x^n). Let $P(x^n)$ be the mariginal distribution of x^n defined by the denominator of eq.(9.1). Then

$$
\begin{aligned}
\mathcal{G}_0(f) &= -\prod_{i=1}^{n} \int dx_i \int dx P(x^n) P(x|x^n) \log f(x|x^n) \\
&= \prod_{i=1}^{n} \int dx_i \int dx P(x^n) P(x|x^n) \log \frac{P(x|x^n)}{f(x|x^n)} \\
&\quad - \prod_{i=1}^{n} \int dx_i \int dx P(x^n) P(x|x^n) \log P(x|x^n).
\end{aligned}
$$

The first term of the right hand side is the average Kullbaclk-Leiber distance from $P(x|x^n)$ to $f(x|x^n)$, and the second term is a constant function of $f(x|x^n)$. Therefore $\mathcal{G}_0(f)$ is minimized if and only if $P(x|x^n) = f(x|x^n)$. $\square$

Remark 71. (1) By this theorem, if we knew the true prior and the true statistical model, there is no statistical inference that attains the smaller average loss than Bayesian predictive inference using the true prior and the true statistical model.

(2) In the real world, we do not know the true distribution. It may seem that the formal optimality theorem does not give any methodology to the real world. However, by the theorem, we can mathematically conclude that the question for finding the optimal statistical inference without information about a prior and a statistical model does not have any answer. In other words, the nature of the statistical estimation is ill-defined, therefore we need the evaluation process of a statistical model and a prior.

Example 62. The same theorems can be proved.

(1) Let $f(x^n)$ be a function from x^n to the parameter space. The error function

$$\mathcal{E}(w) = \int dw \int dx^n P(w, x^n) \|w - f(x^n)\|^2,$$

where $\| \ \|$ is the norm on the parameter space, is minimized if and only if

$$f(x^n) = \frac{\int dw \ w \ P(w, x^n)}{\int dw P(w, x^n)},$$

which is the posterior average of the parameter using the true prior and the true statistical model. If we knew the true prior and the true statistical model, there is no other function which makes the square error smaller. In practical problems, we do not know the true prior and the true statistical model, thus determining the optimal prior and model is the ill-defined problem.

9.2   Bayesian Hypothesis Test

Assume that X_i is an $\mathbb{R}^N$-valued random variable. If a parameter w is subject to a prior $\varphi(w)$ and if $X_1, X_2, ..., X_n$ are independently subject to a probability density $p(x_i|w)$, then such a condition is denoted by

$$w \ \sim \ \varphi(w),$$
$$X_1, X_2, ..., X_n \ \sim \ p(x|w).$$

In this section, we study a Bayesian hypothesis test about a null hypothesis (N.H.) versus an alternative one (A.H.),

$$\text{N.H.} \ : \ w_0 \sim \varphi_0(w_0), \ X_i \sim p_0(x|w_0),$$
$$\text{A.H.} \ : \ w_1 \sim \varphi_1(w_1), \ X_i \sim p_1(x|w_1).$$

Note that w_0 and w_1 may be contained in different sets, for example, $w_0 \in \mathbb{R}^{d_0}$ and $w_1 \in \mathbb{R}^{d_1}$.

Let us study the hypothesis test for the above hypotheses. In order to make a hypothesis test, we need two probabilities for both hypotheses. Let an event Θ be a subset of $\mathbb{R}^{Nn}$. In other words,

$$\Theta \subset \{x^n = (x_1, x_2, ..., x_n) \ ; \ x_i \in \mathbb{R}^N\}.$$

Let Prob($\Theta|N.H.$) and Prob($\Theta|A.H.$) be conditional probabilities of a set Θ where the conditions are defined by the null and alternative hypotheses respectively. The conditional probability density function of $X^n = (X_1, X_2, ..., X_n)$ under the condition that $w \sim \varphi(w), X_i \sim p(x|w)$ is given by

$$\mathcal{P}(x^n|p, \varphi) \equiv \int \varphi(w) \prod_{i=1}^{n} p(x_i|w) dw. \tag{9.3}$$

Then

$$\text{Prob}(\Theta|N.H.) = \int_{\Theta} \mathcal{P}(x^n|p_0, \varphi_0) dx^n, \tag{9.4}$$

$$\text{Prob}(\Theta|A.H.) = \int_{\Theta} \mathcal{P}(x^n|p_1, \varphi_1) dx^n. \tag{9.5}$$

A hypothesis test is defined by an arbitrary pair $(T(x^n), t)$, where $T(x^n)$ is a real-valued function of x^n and t is a real value. Once a hypothesis test is fixed, the decision for a given x^n is determined by

$$\text{If} \quad T(x^n) \leq t \quad \Rightarrow \quad \text{Null hypothesis is chosen,}$$
$$\text{Else} \quad \Rightarrow \quad \text{Alternative hypothesis is chosen.}$$

Any pair $(T(x^n), t)$ gives a test, but we want the better or best one, hence we need an evaluation method of a given test $(T(x^n), t)$. The level and power for a test are respectively defined by

$$\text{Level}(T, t) = \text{Prob}(A.H. \text{ is chosen. } |N.H.), \tag{9.6}$$
$$\text{Power}(T, t) = \text{Prob}(A.H. \text{ is chosen. } |A.H.), \tag{9.7}$$

which are respectively equal to

$$\text{Level}(T, t) = \text{Prob}(T(X^n) > t|N.H.), \tag{9.8}$$
$$\text{Power}(T, t) = \text{Prob}(T(X^n) > t|A.H.). \tag{9.9}$$

That is to say, the level is the probability that A.H. is chosen when N.H. is true, whereas the power is the probability that A.H. is chosen when A.H. is true. A hypothesis test which has a smaller level and a higher power gives the better procedure for decision.

In practical applications, the hypothesis test procedure is conducted as follows.

1. The function for the test $T(x^n)$ is fixed.

2. The level probability is determined. Sometimes 0.05, 0.01, or 0.005 is chosen.

3. For the fixed level probability, the real value t is determined such that Level(T, t) is equal to the level probability.

4. The reject region $\{x^n \; ; \; T(x^n) > t\}$ is determined.

5. If a sample x^n is contained in the reject region, then the alternative hypothesis is chosen, otherwise the null hypothesis is chosen.

If two hypothesis tests $(T(x^n), t)$ and $(U(x^n), u)$ satisfy the condition that

$$\text{Level}(T, t) = \text{Level}(U, u) \Rightarrow \text{Power}(T, t) \geq \text{Power}(U, u),$$

then (T, t) is said to be more powerful than (U, u). This definition gives a partial order on the set of all hypothesis tests. In general, it is not a total order. If there exists a test which is more powerful than any other test, then it is called the most powerful test. In a Bayesian hypothesis test, it is explicitly given by the partition function.

Theorem 28. *Assume that null and alternative hypotheses are given by*

$$N.H. \; : \; w_0 \sim \varphi_0(w_0), \; X_i \sim p_0(x|w_0),$$
$$A.H. \; : \; w_1 \sim \varphi_1(w_1), \; X_i \sim p_1(x|w_1).$$

Then the hypothesis test $(L(x^n), \ell)$ defined by

$$L(x^n) = \frac{\displaystyle\int \varphi_1(w_1) \prod_{i=1}^{n} p_1(x^n|w_1) dw_1}{\displaystyle\int \varphi_0(w_0) \prod_{i=1}^{n} p_0(x^n|w_0) dw_0} \tag{9.10}$$

is the most powerful test.

Proof. Let $(T(x^n), t)$ be an arbitrary hypothesis test. Assume that a real value ℓ is set such that both levels are equal to each other,

$$\text{Level}(L, \ell) = \text{Level}(T, t). \tag{9.11}$$

To prove the theorem, it is sufficient to show

$$\Delta P \equiv \text{Power}(L, \ell) - \text{Power}(T, t)$$

is not smaller than zero. Two events are defined by

$$A = \{x^n \; ; \; T(x^n) - t > 0\},$$
$$B = \{x^n \; ; \; L(x^n) - \ell > 0\}.$$

Here, by the defintion of $L(x^n)$, we can assume $\ell \geq 0$. Then by the definition of eq.(9.8), eq.(9.11) is equivalent to

$$\int_B \mathcal{P}(x^n|p_0, \varphi_0)dx^n - \int_A \mathcal{P}(x^n|p_0, \varphi_0)dx^n = 0. \qquad (9.12)$$

On the other hand by eq.(9.9),

$$\begin{aligned}
\Delta P &= \text{Prob}(L(X^n) > \ell|A.H.) - \text{Prob}(T(X^n) > t|A.H.) \\
&= \int_B \mathcal{P}(x^n|p_1, \varphi_1)dx^n - \int_A \mathcal{P}(x^n|p_1, \varphi_1)dx^n \\
&= \int_{B \cap A^c} \mathcal{P}(x^n|p_1, \varphi_1)dx^n - \int_{A \cap B^c} \mathcal{P}(x^n|p_1, \varphi_1)dx^n,
\end{aligned}$$

where A^c and B^c are complementary sets of A and B respectively. Note that $B \cap A^c \subset B$ and $A \cap B^c \subset B^c$. By eq.(9.10), the condition $x^n \in B$ is equivalent to

$$\int \mathcal{P}(x^n|p_1, \varphi_1)dx^n > \ell \int \mathcal{P}(x^n|p_0, \varphi_0)dx^n.$$

Therefore

$$\begin{aligned}
\Delta P &\geq \ell \int_{B \cap A^c} \mathcal{P}(x^n|p_0, \varphi_0)dx^n - \ell \int_{A \cap B^c} \mathcal{P}(x^n|p_0, \varphi_0)dx^n \\
&= \ell \int_B \mathcal{P}(x^n|p_0, \varphi_0)dx^n - \ell \int_A \mathcal{P}(x^n|p_0, \varphi_0)dx^n = 0,
\end{aligned}$$

where the last equation is derived by eq.(9.12). $\qquad\qquad\square$

Remark 72. For the most powerful test $L(x^n)$, for a given level $\varepsilon > 0$, the reject region $L(x^n) > t$ is determined by choosing t such that

$$\text{Level}(L, t) = \text{Prob}(L(x^n) > t|N.H.) = \varepsilon.$$

Therefore, in order to make a hypothesis test, we need the probability density function of the random variable $L(x^n)$ when the null hypothesis holds.

Example 63. Let us study a case in which a common statistical model is used and two priors are compared,

$$p_0(x|a) \;=\; p_1(x|a) = \frac{1}{\sqrt{2\pi}} \exp(-\frac{1}{2}(x-a)^2), \tag{9.13}$$

$$\varphi_0(a) \;=\; \delta(a), \tag{9.14}$$

$$\varphi_1(a) \;=\; \frac{1}{\sqrt{2\pi}} \exp(-\frac{1}{2}a^2). \tag{9.15}$$

By Theorem 28, the most powerful test is

$$L(x^n) \;=\; \frac{\displaystyle\int \varphi_1(a) \prod_{i=1}^{n} p_1(x^n|a)da}{\displaystyle\int \varphi_0(a) \prod_{i=1}^{n} p_0(x^n|a)da} \tag{9.16}$$

$$=\; \sqrt{\frac{2\pi}{n+1}} \, \exp\Big(\frac{(\sum_{i=1}^{n} x_i)^2}{2(n+1)}\Big). \tag{9.17}$$

Let us make a hypothesis test for the level 0.01. The real value t is determined such that

$$\mathrm{Prob}(L(X^n) > t|N.H.) = 0.01,$$

where the null hypothesis is

$$x^n = (x_1, x_2, ..., x_n) \sim \prod_{i=1}^{n} \frac{1}{\sqrt{2\pi}} \exp(-\frac{1}{2}x_i^2).$$

By eq.(9.17), the condition $L(x^n) > t$ is equivalent to $|(\sum_{i=1}^{n} x_i)/\sqrt{n}| > t^*$, where

$$t^* = \sqrt{(1+1/n)\{2\log t - \log(2\pi/(n+1))\}}.$$

Under the null hypothesis, the random variable $(\sum_{i=1}^{n} X_i)/\sqrt{n}$ is subject to the standard normal distribution $N(x)$. Since

$$\int_{|x|>2.58} N(x)dx = 0.01,$$

the reject region for the level 0.01 is

$$\Big\{x^n \;;\; \Big|\sum_{i=1}^{n} x_i\Big|/\sqrt{n} > 2.58\Big\}.$$

In other words, if $|\sum_{i=1}^{n} x_i|/\sqrt{n} > 2.58$, then the alternative hypothesis is chosen, if otherwise, the null hypothesis is the choice.

9.3 Bayesian Model Comparison

In this section we study a Bayesian model comparison using the posterior distribution. This decision rule is different from the hypothesis test.

Let the prior probabilities a_0 and a_1 satisfy $0 < a_0, a_1 < 1$, $a_0 + a_1 = 1$. We assume that X^n are generated by the following process. Firstly the random variable Y is determined by

$$Y = 0 \quad \text{(with probability } a_0\text{)}, \tag{9.18}$$
$$Y = 1 \quad \text{(with probability } a_1\text{)}. \tag{9.19}$$

Then X^n is generated by

$$\text{If Y=0} \implies w \sim \varphi_0(w), \quad X^n \sim \prod_{i=1}^{n} p_0(x_i|w), \tag{9.20}$$

$$\text{If Y=1} \implies w \sim \varphi_1(w), \quad X^n \sim \prod_{i=1}^{n} p_1(x_i|w). \tag{9.21}$$

In this case the simultaneous probability density function of (Y, X^n) is given by

$$p(y, x^n) = \Big(a_0 \mathcal{P}(x^n|\varphi_0, p_0)\Big)^{1-y} \Big(a_1 \mathcal{P}(x^n|\varphi_1, p_1)\Big)^{y}, \tag{9.22}$$

where the definition of $\mathcal{P}(x^n|\varphi, p)$ is given in eq.(9.3). The posterior probability of $Y = 1$ for a given X^n is

$$p(1|x^n) = \frac{p(1, x^n)}{p(0, x^n) + p(1, x^n)}, \tag{9.23}$$

which gives the posterior model decision. Let us study the decision rule by which the random variable Y is estimated by the following random variable Z,

$$p(1|x^n) > \alpha \implies Z = 1, \tag{9.24}$$
$$p(1|x^n) \leq \alpha \implies Z = 0, \tag{9.25}$$

where $0 < \alpha < 1$ is a constant. Then the condition $p(1|x^n) > \alpha$ is equivalent to

$$\frac{p(1|x^n)}{p(0|x^n)} > \frac{1 - \alpha}{\alpha}.$$

By eq.(9.22), it is equivalent to

$$\frac{a_1 \mathcal{P}(x^n|\varphi_1, p_1)}{a_0 \mathcal{P}(x^n|\varphi_0, p_0)} > \frac{1 - \alpha}{\alpha}.$$

By eq.(9.3), also it is equivalent to

$$
\frac{\displaystyle\int \varphi_1(w_1) \prod_{i=1}^{n} p_1(x^n|w_1)dw_1}{\displaystyle\int \varphi_0(w_0) \prod_{i=1}^{n} p_0(x^n|w_0)dw_0} > \frac{a_0}{a_1} \cdot \frac{1-\alpha}{\alpha}.
\tag{9.26}
$$

This inequality would be equal to the most powerful test if $a_0(1-\alpha)/(a_1\alpha) = t$. Note that, if $a_0 = a_1 = \alpha = 1/2$, then the right hand side of this inequality is equal to 1.

Example 64. Let us compare the most powerful test with the posterior model comparison. Let us adopt the same case as Example 63,

$$
p_0(x|a) = p_1(x|a) = \frac{1}{\sqrt{2\pi}} \exp(-\frac{1}{2}(x-a)^2),
\tag{9.27}
$$

$$
\varphi_0(a) = \delta(a),
\tag{9.28}
$$

$$
\varphi_1(a) = \frac{1}{\sqrt{2\pi}} \exp(-\frac{1}{2}a^2).
\tag{9.29}
$$

Then by eq.(9.17),

$$
\sqrt{\frac{2\pi}{n+1}} \exp\left(\frac{(\sum_{i=1}^{n} x_i)^2}{2(n+1)}\right) > \frac{a_0}{a_1} \cdot \frac{1-\alpha}{\alpha}.
$$

Hence

$$
\frac{|\sum_{i=1}^{n} x_i|}{\sqrt{n}} > \left[\frac{n+1}{n}\log\left(\frac{n+1}{2\pi}\right) + \frac{2(n+1)}{n}\log\left(\frac{a_0}{a_1} \cdot \frac{1-\alpha}{\alpha}\right)\right]^{1/2}
$$

$$
\cong \sqrt{\log n}.
\tag{9.30}
$$

If n is sufficiently large, then the right hand side is approximated by $\sqrt{\log n}$, whereas the most powerful test for the level 0.01 is 2.58. That is to say, in the hypothesis test

$$
\frac{|\sum_{i=1}^{n} x_i|}{\sqrt{n}} < 2.58 \Longrightarrow \text{Model 0 is chosen,}
$$

whereas in the posterior model comparison,

$$
\frac{|\sum_{i=1}^{n} x_i|}{\sqrt{n}} < \sqrt{\log n} \Longrightarrow \text{Model 0 is chosen.}
$$

There is no mathematical contradiction because the hypothesis test and the posterior model comparison are different procedures based on different assumptions. However, this difference is sometimes referred to as a paradox. The former has the same decision order as the model selection by the cross validation or information criteria, whereas the latter does that by the free energy.

9.4 Phase Transition

Let us define a phase transition in Bayesian statistics.

Definition 29. (Phase transition) If a statistical model $p(x|w)$ or a prior $\varphi(w)$ is determined by a value θ which is not the parameter, then it is written as $p(x|w,\theta)$ or $\varphi(w|\theta)$. Therefore, the posterior distribution is also written as $p(w|X^n,\theta)$. Such a value θ is called a generalized hyperparameter. If a posterior distribution for a sufficiently large n changes drastically at $\theta = \theta_c$, then it is said that the posterior distribution has a phase transition, and θ_c is called a critical point. At a critical point, the free energy $F_n(\beta,\theta)$ is often discontinuous or nondifferentiable.

If a log density ratio function $f(x,w) = \log(q(x)/p(x|w))$ has relatively finite variance, then

$$F_n - S_n = -\log \int \exp\Big(-\sum_{i=1}^n f(X_i,w)\Big)\varphi(w)dw$$

and

$$\overline{F}_n - S = -\log \int \exp\Big(-n\mathbb{E}_X[f(X,w)]\Big)\varphi(w)dw$$

have the same asymptotic expansions according to the order that is larger than $O(1)$. Hence we can analyze the phase transition by studying $\overline{F}_n$ instead of F_n.

Example 65. Firstly, we study a case when the log density ratio function has a generalized hyperparameter. Let $w = (x,y)$ and

$$\overline{F}_n(\theta) = -\log \int_0^1 dx \int_0^1 dy \, \exp(-nx^2 y^\theta).$$

Let $K(x,y) = x^2 y^\theta$ $(\theta > 0)$. Then $K(x,y) \geq 0$ and the set of all zero points of $K(x,y)$ is $\{(x,y); xy = 0\}$. It can be rewritten as

$$\overline{F}_n(\theta) = -\log \int_0^n \frac{dt}{n} \int_0^1 dx \int_0^1 dy \, \exp(-t) \, \delta(t/n - K(x,y)).$$

Since $K(x, y)$ is a normal crossing function whose multi-indexes are $k = (2, \theta)$ and $h = (0, 0)$, the behavior of the free energy can be analyzed by using the zeta function.

$$\zeta(z) = \int_0^1 dx \int_0^1 dy (x^2 y^\theta)^z = \frac{1}{(2z + 1)(\theta z + 1)}.$$

Hence the state density function is equal to

$$\delta(t - K(x, y)) \cong \begin{cases} 1/(2\theta) \cdot t^{-1/2} \delta(x) y^{-\theta/2} & (\theta < 2) \\ 1/4 \cdot t^{-1/2}(-\log t)\delta(x)\delta(y) & (\theta = 2) \\ 1/(2\theta) \cdot t^{1/\theta - 1} x^{-2/\theta} \delta(y) & (\theta > 2). \end{cases}$$

Therefore, the critical point is $\theta = 2$, where the posterior distribution drastically changes from $\delta(x)$ to $\delta(y)$ between $(2 - \epsilon) \to (2 + \epsilon)$. The asymptotic behavior of the free energy is given by

$$\overline{F}_n(\theta) \cong \begin{cases} (1/2) \log n + O(1) & (\theta < 2) \\ (1/2) \log n - \log \log n + O(1) & (\theta = 2) \\ (1/\theta) \log n + O(1) & (\theta > 2) \end{cases},$$

which shows that the coefficient of $\log n$ is a continuous function of θ but not differentiable at $\theta = 2$.

In Bayesian statistics, the following concepts are all mathematically connected. In order to analyze the phase transition, we can choose the most convenient one.
(1) Partition function
(2) Free energy or the minus log marginal likelihood
(4) State density function
(5) Zeta function
(6) Posterior distribution

Example 66. Secondly, let us study a case when the prior has a hyperparameter. Assume that $w = (x, y)$, $x \in \mathbb{R}^1$, and $y \in \mathbb{R}^M$. Let us study a free energy,

$$\overline{F}_n(\theta) = - \log \int_0^1 dx \int_{\|y\| < 1} dy \, \exp(-nx^2 \|y\|^2) x^{\theta - 1}.$$

The zeta function is equal to

$$\zeta(z) = \int_0^1 dx \int_{\|y\| < 1} dy \, x^{2z + \theta - 1} \|y\|^{2z}.$$

By using the generalized polar system (r, ψ) such that $y = r\psi$, then $dy = r^{M-1}drd\psi$, resulting that

$$\zeta(z) = \int_0^1 dx \int_0^1 dr \int d\psi \, x^{2z+\theta-1} r^{2z+M-1}$$

$$= \frac{1}{(2z+\theta)(2z+M)}\left(\int d\psi\right).$$

Hence by using $\Psi = \int d\psi$, it follows that the state density function is

$$\delta(t - x^2\|y\|^2)x^{\theta-1} \cong \begin{cases} (1/4)\Psi \cdot t^{\theta/2-1}\delta(x)y^{M-\theta-1} & (\theta < M) \\ (1/4)\Psi \cdot t^{M/2-1}x^{\theta-M-1}\delta(y) & (\theta > M). \end{cases}$$

Therefore the posterior distribution drastically changes from $\delta(x)$ to $\delta(y)$ at the critical point $\theta = M$. Then the free energy is

$$\overline{F}_n(\theta) \cong \begin{cases} (\theta/2)\log n + O(1) & (\theta < M) \\ (M/2)\log n + O(1) & (\theta > M) \end{cases},$$

which shows that the coefficient of $\log n$ of the free energy is continuous at $\theta = M$ but not differentiable.

Example 67. Let us study a normal mixture of $x \in \mathbb{R}^2$ for a given parameter a $(0 \le a \le 1)$ and $b \in \mathbb{R}^2$,

$$p(x|a, b) = aN(x|0) + (1-a)N(x|b).$$

For a prior, we use the Dirichlet distribution with index $\alpha > 0$,

$$\varphi(a) \propto (a(1-a))^{\alpha-1},$$

$$\varphi(b) \propto \exp(-\frac{\|b\|^2}{2\sigma^2}),$$

where α and σ are hyperparameters. We set $\sigma = 10$, and study the phase transition according to the hyperparameter α. Assume that the true distribution is

$$q(x) = N(x|0).$$

Then the zeta function is

$$\zeta(z) = \int K(a, b)^z \varphi(a)\varphi(b)dadb,$$

where

$$K(a, b) = \int q(x) \log \frac{q(x)}{p(x|a, b)}dx.$$

Then in the neighborhood of $(a,b) = 0$, there exists $c_1, c_2 > 0$ such that

$$c_1 a^2 \|b\|^2 < K(a,b) < c_2 a^2 \|b\|^2.$$

Hence the real log canonical threshold is

$$\lambda = \min\{\alpha/2, 1\},$$

which shows that there exists a phase transition at the critical point $\alpha = 2$. Thus the average generalization and cross validation errors are given by

$$
\begin{aligned}
\mathbb{E}[G_n] &= \min\{\alpha/2, 1\}/n + o(1/n), \\
\mathbb{E}[C_n] &= \min\{\alpha/2, 1\}/n + o(1/n).
\end{aligned}
$$

Moreover, if $\alpha < 2$, then a is in the neighborhood of the origin but b is free, whereas, if $\alpha > 2$, then b is in the neighborhood of the origin but a is free. Therefore, the posterior distribution drastically changes at the critical point. Moreover, at the critical point $\alpha = 2$, the posterior distribution is unstable, hence MCMC processes may have large variance. Let us observe the phase transition by an experiment. The number of independent random variables was set as $n = 100$. The hyperparameters α of the prior distribution were controlled in $0 < \alpha < 6$. Figures 9.1 and 9.2 show the distributions of the generalization errors and cross validation errors for a given hyperparameter respectively. The circles in both figures show their averages. Note that, by the equation

$$(G_n - S) + (C_n - S_n) = \frac{2\lambda}{n} + o_p(1/n),$$

if $G_n - S$ is larger than the average, then $C_n - S_n$ is smaller than the average.

Remark 73. (1) In this section, we study the cases where we can derive the poles of the zeta functions. In general, in order to find them, resolution of singularities is necessary, which may often be difficult. That is to say, it is not easy to find the critical point rigorously.

(2) If we apply the mean field approximation, or equivalently the variational Bayes, to statistical estimation, then the posterior distribution made by the mean field approximation shrinks or becomes localized. As a result, the free energy becomes larger, and the phase transition structure changes. Sometimes a spurious phase transition can be observed in the mean field approximation which does not exist in the true posterior distribution. The critical points of the mean field approximation and the true posterior do not coincide in general.

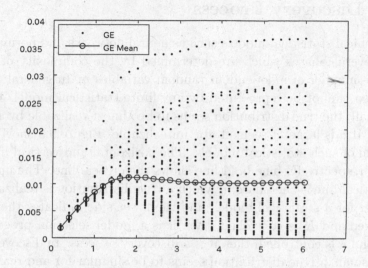

Figure 9.1: Phase transition of generalization error in a normal mixture. The horizontal and vertical lines show Dirichlet hyperparameter α and the generalization error respectivelly. $\alpha = 2$ is the critical point.

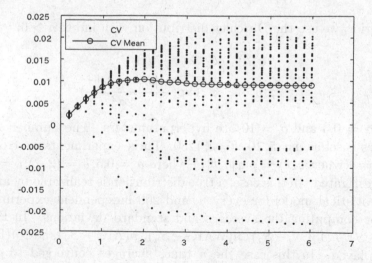

Figure 9.2: Phase transition of cross validation error in a normal mixture. The horizontal and vertical lines show Dirichlet hyperparameter α and the cross validation error respectivelly. $\alpha = 2$ is the critical point.

9.5 Discovery Process

Hierarchical statistical models such as neural networks and normal mixtures have several phases which are determined by the complexity of models K and the number of independent random variables n. In general, a true distribution cannot be represented by any finite statistical model. However, if n is small, the true distribution seems to be almost realizable by a statistical model. If n is large, it seems to be unrealizable. Figure 9.3 shows the phase diagram of such statistical models. The horizontal and vertical lines show n and K respectively. The bold line shows the critical line. The upper side of the critical line is the phase in which a true distribution is realizable by and singular for a statistical model, and the lower side indicates the reverse. If n is fixed and K is controlled, then it is a model selection process. If K is fixed and n is controlled, then it is a discovery process. In discovery process, if n is small, a true distribution seems to be singular for and realizable by a statistical model, otherwise, it seems to be regular and unrealizable. In this section we study a discovery process.

Let us consider a normal mixture of $x \in \mathbb{R}^2$ for a given parameter a $(0 \le a \le 1)$ and $b, c \in \mathbb{R}^2$,

$$p(x|a, b, c) = aN(x|b) + (1 - a)N(x|c).$$

For a prior, we use the Dirichlet distribution with index $\alpha > 0$,

$$\varphi(a) \ \propto \ (a(1 - a))^{\alpha - 1},$$
$$\varphi(b, c) \ \propto \ \exp(-\frac{\|b\|^2 + \|c\|^2}{2\sigma^2})$$

where $\alpha = 0.3$ and $\sigma = 10$ are hyperparameters. The number of random variables is set as $n = 5, 10, 20, ..., 1280$. Three experiments were conducted. (1) A case when $q(x) = p(x|a, b, c)$ where $a = 0.5, b = (2, 2), c = (-2, -2)$ was investigated. In this case, a true distribution is realizable by and regular for a statistical model for every n, and 200 independent experiments were used for computing the averages and standard deviations. In Figure 9.4, $n(G_n - S)$, $n(\text{ISCV} - S_n)$, $n(\text{WAIC} - S_n)$, $n(\text{AIC} - S_n)$, and $n(\text{DIC} - S_n)$ are displayed. In this case the n times averages converged to $d/2 = 2.5$ where d is the dimenson of the parameter. If $n \ge 20$, every criterion could estimate the generalization error. In this case, there was no phase transition. (2) A case when $q(x) = p(x|a, b, c)$ where $a = 0.5, b = (0, 0), c = (0, 0)$ was investigated. In this case, a true distribution was realizable by but nonregular for a statistical model for every n. In this case the n times average of

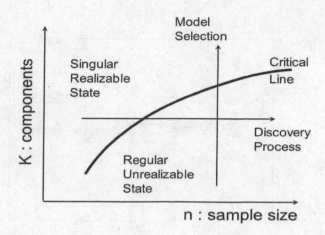

Figure 9.3: Phase diagram. Statistical models such as neural networks and normal mixtures have phase transitions according to the number of components and the sample size. If n is small, then a true distribution seems to be realizable and singular. If n is large, then it seems to be unrealizable and regular.

the generalization error converged to $\lambda = 1.5$ where λ is the real log canonical threshold. In Figure 9.5, $n(G_n - S)$, $n(\text{ISCV} - S_n)$, $n(\text{WAIC} - S_n)$, $n(\text{AIC} - S_n)$, and $n(\text{DIC} - S_n)$ are displayed, which shows that both ISCV and WAIC could estimate the generalization losses, whereas neither AIC nor DIC could. In this case, there was no phase transition.

(3) A case when $q(x) = p(x|a, b, c)$ where $a = 0.5, b = (0.5, 0.5), c = (-0.5, -0.5)$ was investigated. In Figure 9.6, $n(G_n - S)$, $n(\text{ISCV} - S_n)$, $n(\text{WAIC} - S_n)$, $n(\text{AIC} - S_n)$, and $n(\text{DIC} - S_n)$ are displayed. In the region $n \leq 20$, the generalization loss was almost equal to the case when the true distribution is singular for a statistical model, which is the case (1). In the region $n \geq 320$, it was almost equal to the case when the true distribution is regular for a statistical model, which is the case (2). And in the region $40 \leq n \leq 160$, the generalization errors were larger than in the other cases, which were on the critical point. Both ISCV and WAIC could estimate the generalization losses, whereas neither AIC nor DIC could. In this case, there was a phase transition according to knowledge discovery process.

By using the same model as above (3), let us study the statistical estimation problem of the hidden variable. A hidden variable y is introduced

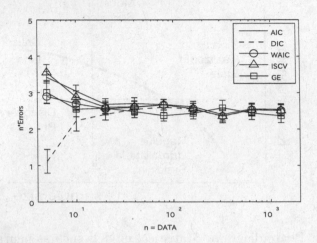

Figure 9.4: Strict regular case. The horizontal and vertical lines show the sample size and errors respectively. A true distribution is set as realizable by and regular for a statistical model. No phase transition occurs.

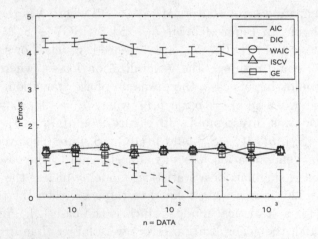

Figure 9.5: Strict singular case. The horizontal and vertical lines show the sample size and errors respectively. A true distribution is set as realizable by and singular for a statistical model. No phase transition occurs.

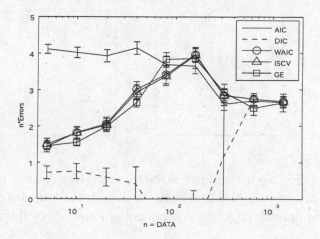

Figure 9.6: Discovery process. The horizontal and vertical lines show the sample size and errors respectively. As n increases the true structure is discovered by a statistical model. A phase transition occurs.

by the following equation,

$$p(x, y|a, b, c) = [aN(x|b)]^y \cdot [(1-a)N(x|c)]^{1-y}.$$

That is to say,

$$
\begin{aligned}
p(x, 0|a, b, c) &= aN(x - b), \\
p(x, 1|a, b, c) &= (1-a)N(x - c),
\end{aligned}
$$

By marginalizing y this model results in a normal mixture,

$$p(x|a, b, c) = aN(x - b) + (1-a)N(x - c).$$

The likelihood function of (x^n, y^n) is given by

$$p(x^n, y^n|a, b, c) = \prod_{i=1}^{n} [aN(x_i - b)]^{y_i} \cdot [(1-a)N(x_i - c)]^{1-y_i}.$$

Figures 9.7, 9.8, and 9.9 show the true distribution $q(x)$ at left, and estimated hidden variables for $n = 10, 100, 1000$ at right. If the hidden variable of a sample point x_i was estimated $y_i < 0.5$, x_i is shown by a point, if otherwise, a white square. For $n = 10$, it seems that a true distribution consists of one normal distribution, whereas for $n = 1000$, two distributions. For the case $n = 100$, it is on a critical point, hence the estimated results were unstable.

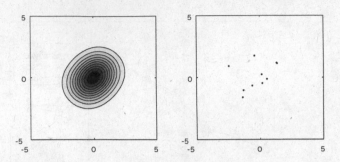

Figure 9.7: Estimated hidden variables $n = 10$. A true distribution and estimated latent variables in $n = 10$. Although the true distribution consists of two components, only one component is found since n is too small.

Remark 74. (1) In the above experiment, the hidden variable $y = (0, 1)$ can be replaced by $y = (1, 0)$, because of symmetry. However, MCMC naturally made this symmetry break down, which was used for distinguishing 0 and 1.

(2) The phase transition affects the estimation of hidden variables. The phase transition also affects the MCMC process. Hierarchical statistical models such as neural networks and normal mixtures have the same phase transition as this case, hence a statistician must know its structure before applying statistical models to the real world problems.

9.6 Hierarchical Bayes

In this section, hierarchical Bayesian estimation is studied. A typical case is explained by using an example.

Example 68. (Hierarchical Bayesian inference) In a high school, there are $m = 10$ classes, and each class has $n = 30$ students. One day, an examination of mathematics was done, and $mn = 300$ scores $\{x_{ki}; 1 \le k \le m, 1 \le i \le n\}$ were obtained. We would like to analyze the following statistical model.

(1) The average w_k of kth class' scores are subject $N(\mu, 1^2)$.

(2) The score x_{ki} of the ith student in the kth class is subject to $N(w_k, 10^2)$. That is to say, the model is made by

$$
\begin{aligned}
w_k &\sim N(\mu, 1^2), \\
x_{ki} &\sim N(w_k, 10^2).
\end{aligned}
$$

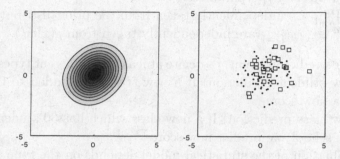

Figure 9.8: Estimated hidden variables $n = 100$. A true distribution and estimated latent variables in $n = 100$. The posterior distribution is unstable, because it lies on the critical point between one component and two components.

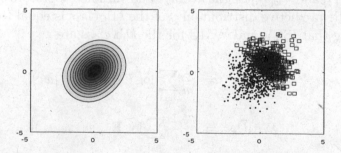

Figure 9.9: Estimated hidden variables $n = 1000$. A true distribution and estimated latent variables in $n = 1000$. Two components were found.

We would like to evaluate this model according to the obtained data.

Statistical Model
(1) Let μ be the hyperparameter.
(2) $\{w_k\}_{k=1}^m$ are independently taken from the prior distribution $\varphi(w|\mu)$.
(3) $(x_k)^n \equiv \{x_{ki}\}_{i=1}^n$ are independently taken from $p(x|w_k)$.

From the predictive point, there are at least two different types of prediction.
(1) (New student prediction) If a new student is added to the k class, we predict a new score.
(2) (New class prediction) If a new class which has 30 students is added to the high school, we predict new scores.
The evaluation of the statistical model depends on the type of prediction.

New Student Prediction. For given all data $\{(x_k)^n\}$, the posterior distribution of all classes $(w_1, w_2, ..., w_m)$ is

$$p(w_1, w_2, ..., w_m|(x_1)^n, (x_2)^n, ..., (x_m)^n) \propto \prod_{k=1}^m \left(\varphi(w_k|\mu) \prod_{i=1}^n p(x_{ki}|w_k) \right).$$

This distribution shows that $(w_1, w_2, ..., w_m)$ are independent of each other, hence

$$p(w_k|x_k^n) \propto \varphi(w_k|\mu) \prod_{i=1}^n p(x_{ki}|w_k).$$

Let $\mathbb{E}_{w_k}[\ ]$ and $\mathbb{V}_{w_k}[\ ]$ be the average and variance operators of this distribution. The predictive distribution y of the kth class is equal to $\mathbb{E}_{w_k}[p(y|w_k)]$, resulting that ISCV and WAIC for the kth class are

$$\text{ISCV}_k = \frac{1}{n}\sum_{i=1}^n \log \mathbb{E}_{w_k}[1/p(x_{ik}|w_k)], \qquad (9.31)$$

$$\text{WAIC}_k = -\frac{1}{n}\sum_{i=1}^n \log \mathbb{E}_{w_k}[p(x_{ik}|w_k)]$$

$$+\frac{1}{n}\sum_{i=1}^n \mathbb{V}_{w_k}[\log p(x_{ik}|w_k)]. \qquad (9.32)$$

If a student is added to every class, then

$$\sum_k \text{ISCV}_k, \quad \sum_k \text{WAIC}_k$$

are equal to ISCV and WAIC, respectively. Therefore, the model evaluation can be done by using this value.

New Class Prediction. In the second problem, one sample point is $(x_k)^n$ and we have a sample which consists of $(x_1)^n$, $(x_2)^n$, ...,$(x_m)^n$. The statistical model is

$$\mathcal{P}((x_k)^n|\mu) = \int \varphi(w|\mu) \prod_{i=1}^{n} p(x_{ki}|w)dw.$$

In this case, the hyperparameter is parameter of this model. By setting a posterior distribution $\psi(\mu)$ of μ, the posterior distribution is given by

$$p(\mu|(x_1)^n, (x_2)^n, ..., (x_m)^n) \propto \psi(\mu) \prod_{k=1}^{m} \mathcal{P}((x_k)^n|\mu).$$

Let $\mathbb{E}_\mu[\ ]$ and $\mathbb{V}_\mu[\ ]$ be average and variance operators by this distribution. The predictive distribution of y^n is equal to $\mathbb{E}_\mu[\mathcal{P}(y^n|\mu)]$, resulting that

$$\text{ISCV} = \frac{1}{m} \sum_{k=1}^{m} \log \mathbb{E}_\mu[1/\mathcal{P}((x_k)^n|\mu)], \qquad (9.33)$$

$$\text{WAIC} = -\frac{1}{m} \sum_{k=1}^{m} \log \mathbb{E}_\mu[\mathcal{P}((x_k)^n|\mu)]$$

$$+ \frac{1}{m} \sum_{k=1}^{m} \mathbb{V}_\mu[\log \mathcal{P}((x_k)^n|\mu)]. \qquad (9.34)$$

In the second problem, μ is estimated by the posterior distribution. In general, it is rather difficult to numerically calculate the cross validation and WAIC in the second case.

Remark 75. Cross validation and information criteria are defined for evaluation of the predictive behavior of the statistical estimation. A complex statistical model such as hierarchical Bayes methods may yield several different predictions. To make an evaluation method, a statistician should determine which prediction would be evaluated.

Example 69. (Hierarchical Bayes in linear regression) Assume that $x, y \in \mathbb{R}$ and $a_k, b_k, s \in \mathbb{R}$. Let K be the number of the groups. For each k, a statistical model $p(y|x, a_k, b_k, s)$ and a prior $\varphi(a_k, b_k|m_a, m_b, t)\phi(s|r, \varepsilon)$ are defined by

$$p(y|x, a_k, b_k, s) = \sqrt{\frac{s}{2\pi}} \exp(-\frac{s}{2}(y - a_k x - b_k)^2),$$

$$\varphi(a_k, b_k|m_a, m_b) = \frac{t}{2\pi} \exp(-\frac{t}{2}\{(a_k - m_a)^2 + (b_k - m_b)^2\}),$$

$$\phi(s|r) = \frac{\varepsilon^r}{\Gamma(r)} s^{r-1} \exp(-\varepsilon s),$$

where (m_a, m_b) and r are hyperparameters. Note that (m_a, m_b) is the common average of the prior which is estimated by using the uniform and improper prior, and r is optimized by the cross validation and WAIC. Let $t = 1/0.2^2$ and $\varepsilon = 0.01$ be fixed. Assume that the true distribution is given by the common conditional density $p(y|x, a_0, b_0, s_0)$ where $a_0 = 1$, $b_0 = 0$, and $s_0 = 0.1$. Let $\{(x_{ki}, y_{ki})\}_{i=1}^{n_k}$ be a sample for the kth group whose sample size is n_k. The posterior distribution for (s, a_k, b_k, m_a, m_b) is in proportion to

$$\phi(s|r) \prod_{k=1}^{K} \varphi(a_k, b_k|m_a, m_b) \prod_{i=1}^{n_k} p(y_{ki}|x_{ki}, a_k, b_k, s),$$

which is also in proportion to

$$s^{r-1} \exp(-\varepsilon s) \prod_{k=1}^{K} \frac{t}{2\pi} \exp(-\frac{t}{2}\{(a_k - m_a)^2 + (b_k - m_b)^2\})$$

$$\times \prod_{i=1}^{n_k} \sqrt{\frac{s}{2\pi}} \exp(-\frac{s}{2}(y_{ki} - a_k x_{ki} - b_k)^2).$$

This posterior density can be approximated by a Gibbs sampler,

$$p(m_a|s, a_k, b_k) = \mathcal{N}(\,(1/K)\sum_{k=1}^{K} a_k, \ \sqrt{1/(tK)}\,),$$

$$p(m_b|s, a_k, b_k) = \mathcal{N}(\,(1/K)\sum_{k=1}^{K} b_k, \ \sqrt{1/(tK)}\,),$$

$$p(s|a_k, b_k) = \mathcal{G}(\,r + (1/2)\sum_{k=1}^{K} n_k, \ 1/B),$$

$$p(a_k, b_k|m_a, m_b, s) = \mathcal{N}(A^{-1}v, A^{-1/2}),$$

r	-3	-2	-1	0	1	2	3	4
CV	0.96	0.87	0.77	0.70	0.66	0.66	0.70	0.76
WAIC	0.61	0.53	0.46	0.43	0.42	0.45	0.51	0.58

Table 9.1: Hyperparameters, cross validation error, and WAIC error. Cross validation and WAIC errors are compared as a function of hyperparameter in hierarchical Bayes. In this case $r = 1$ is chosen which minimizes both cross validaiton and WAIC errors.

(Table)

where $\mathcal{N}(m, S)$ is the normal distribution whose average and covariance are m and S, respectively and $\mathcal{G}(a, b)$ is the gamma distribution defined by eq.(8.74), and

$$A = \begin{pmatrix} (t + s\sum_{i=1}^{n_k} x_{ki}^2) & (s\sum_{i=1}^{n_k} x_{ki}) \\ (s\sum_{i=1}^{n_k} x_{ki}) & (t + sn_k) \end{pmatrix}$$

$$v = \begin{pmatrix} tm_a + s\sum_{i=1}^{n_k} y_{ki}x_{ki} \\ tm_b + s\sum_{i=1}^{n_k} y_{ki} \end{pmatrix}$$

$$B = \varepsilon + (1/2)\sum_{k=1}^{K}\sum_{i=1}^{n_k}(y_{ki} - a_k x_{ki} - b_k)^2\}$$

For $K = 6$, $n_k = 5 + k$, the cross validation and WAIC errors for the first case are compared. See Table 69. In this case, $r = 1$ which minimizes both errors is chosen as the best hyperparameter.

9.7 Problems

1. Assume that w is taken from $\Phi(w)$, and then (X^{n+1}, Y^{n+1}) are independently taken from $P(x, y|w)$. For function $f(x|x^n, y^n)$ from x to y, which is determined by (x^n, y^n), the square error is defined by

$$\mathcal{E}(f) = \int dw\Phi(w)\int dx^{n+1}dy^{n+1}P(x^{n+1}, y^{n+1}|w)$$
$$\times \|y_{n+1} - f(x_{n+1}|x^n, y^n)\|^2,$$

where $\|y\|$ is the norm of y. Prove that this square error is minimized if and only if

$$f(x_{n+1}|x^n, y^n) = \frac{\int dw\Phi(w)\int dy_{n+1}\, y_{n+1}\, P(x^{n+1}, y^{n+1}|w)}{\int dw\Phi(w)\int dy_{n+1}P(x^{n+1}, y^{n+1}|w)},$$

which is the regression function made from the predictive distribution.

2. Let $x, a \in \mathbb{R}^d$. A statistical model and a prior are defined by

$$p(x|a) = \frac{1}{(2\pi)^{d/2}} \exp(-\frac{1}{2}\|x - a\|^2),$$

$$\varphi(a) = \frac{1}{(2\pi)^{d/2}} \exp(-\frac{1}{2}\|a\|^2).$$

Devise a hypothesis test for the null hypothesis $(p(x|a), \delta(a))$ versus the alternative one $(p(x|a), \varphi(a))$. Then compare it with the minimum cross validation loss estimation defined by eq.(8.75).

3. Let $x, b \in \mathbb{R}^d$ and $y, a \in \mathbb{R}$. For a statistical model and a prior,

$$p(y|x, a, b) = \frac{1}{(2\pi)^{d/2}} \exp(-\frac{1}{2}\|y - a\tanh(b \cdot x)\|^2),$$

$$\varphi(a, b) \propto |a|^{\alpha - 1},$$

clarify the phase transition structure according to the hyperparameter $\alpha > 0$.

4. Let $a = (a_1, a_2, ..., a_K)$ satisfy $\sum a_k = 1$ and $a_k \geq 0$. Also let $b_k \in \mathbb{R}$. A statistical model of $x \in \mathbb{R}^N$

$$p(x|a, b) = \sum_{k=1}^{K} \frac{a_k}{\sqrt{2\pi}} \exp(-\frac{(x - b_k)^2}{2})$$

is called a normal mixture. Let the prior be a constant on (a, b). Assume that a true distribution is given by

$$q(x) = \sum_{k=1}^{\infty} \frac{1}{2^k} \exp\left(-\frac{(x - k)^2}{2}\right).$$

Then explain the discovery process of this case.

Chapter 10

Basic Probability Theory

In this chapter, we summarize the basic probability theory which is an important component in this book.

(1) Delta function is defined.

(2) Kullback-Leibler distance is introduced and its mathematical property is proved.

(3) The definitions of the probability space and the random variable are described.

(4) An empirical process is a random variable on a function space. In Bayesian theory construction, we need its convergence in distribution. The basic definitions and essential theorems are introduced.

(5) Even if a sequence of random variables converges in distribution, the sequence of its expected values may not converge. To prove the convergence of the expected values, we need the additional condition such as the asymptotically uniformly integrablility.

10.1 Delta Function

The delta function $\delta(x)$ is defined to be a generalized function of $x \in \mathbb{R}$ which satisfies, for an arbitrary continuous function $f(x)$,

$$\int \delta(x)f(x)dx = f(0).$$

The delta function is not an ordinary function but a kind of a distribution, which formally satisfies

$$\delta(0) = \infty, \quad \int \delta(x)dx = 1.$$

We can understand that $\delta(x)$ is the probability density function of the random variable X that is $X = 0$ almost surely. For $x = (x_1, x_2, ..., x_N) \in \mathbb{R}^N$, the multi-dimensional delta function $\delta(x)$ is defined by

$$\delta(x) = \delta(x_1)\delta(x_2)\cdots\delta(x_N).$$

Example 70. Assume $a > 0$. Then

$$a\delta(ax + b) = \delta(x + b/a). \tag{10.1}$$

Let us show this equality. Let $f(x)$ be an arbitrary continuous function. By the transform $y = ax + b$, $dy = adx$ and

$$\int a\delta(ax + b)f(x)dx = \int \delta(y)f((y - b)/a)dy = f(-b/a).$$

On the other hand,

$$\int \delta(x + b/a)f(x)dx = f(-b/a).$$

10.2   Kullback-Leibler Distance

Let $q(x)$ and $p(x)$ be probability density functions on $\mathbb{R}^N$. Then the Kullback-Leibler distance or the relative entropy is defined by

$$D(q\|p) = \int q(x)\log\frac{q(x)}{p(x)}dx.$$

The Kullback-Leibler distance indicates the difference between $q(x)$ and $p(x)$ by the following lemma.

Lemma 26. *Assume that $q(x)$ and $p(x)$ are continuous functions.*

  1. For arbitrary $q(x)$ and $p(x)$, $D(q\|p) \geq 0$.

  2. $D(q\|p) = 0$ if and only if $q(x) = p(x)$ for all x such that $q(x) > 0$.

Proof. Let us define a function $U(t)$ for $t > 0$ by

$$U(t) = 1/t + \log t - 1.$$

Then

$$D(q\|p) = \int q(x)U\left(\frac{q(x)}{p(x)}\right)dx.$$

Note that $U(t) \geq 0$ for arbitrary $t > 0$. Since $q(x) \geq 0$, the first half of thelemma was proved. Let us prove the second half. If $q(x) = p(x)$ for arbitrary x such that $q(x) > 0$, then $D(q||p) = 0$. Assume $D(q||p) = 0$. $U(t)$ is a continuous function of t, hence $U(q(x)/p(x))$ is a continuous function of x. Thus $U(q(x)/p(x)) = 0$ for arbitrary x such that $q(x) > 0$. Hence $q(x) = p(x)$, because $U(t) = 0$ is equivalent to $t = 1$. $\square$

The log loss function of $p(x)$ is defined by

$$L(p) = - \int q(x) \log p(x) dx.$$

Then by the definition of Kullback-Leibler distance,

$$
\begin{aligned}
L(p) &= \int q(x) \log(q(x)/p(x)) dx - \int q(x) \log q(x) dx \\
&= D(q||p) + S.
\end{aligned}
\tag{10.2}
$$

In this equation, S is the entropy of $q(x)$ which does not depend on $p(x)$ and $D(q||p) \geq 0$. That is to say,

$$L(p) \text{ is small} \iff D(q||p) \text{ is small}. \tag{10.3}$$

In other words, minimization of KL distance is equivalent to minimization of $L(p)$.

Remark 76. (Calculation of generalization error) As is shown in the above proof,

$$\int q(x) \log \frac{q(x)}{p(x)} dx = \int q(x) U\left(\frac{q(x)}{p(x)}\right) dx,$$

where $U(x) \geq 0$. There are two different ways to approximate the Kullback-Leibler distance. Let $\{X_i\}$ be a set of random variables which are independently subject to the probability density function $q(x)$. Then $D(q||p)$ can be approximated by

$$D_1 \approx \frac{1}{n} \sum_{i=1}^{n} \log \frac{q(X_i)}{p(X_i)},$$

$$D_2 \approx \frac{1}{n} \sum_{i=1}^{n} U\left(\frac{q(X_i)}{p(X_i)}\right).$$

Sometimes variance of D_2 is smaller than D_1.

10.3 Probability Space

Definition 30. (Metric space) Let Ω be a set. A function D

$$D : \Omega \times \Omega \ni (x, y) \mapsto D(x, y) \in \mathbb{R}$$

is called a metric if it satisfies the following three conditions.
(1) For arbitrary $x, y \in \Omega$, $D(x, y) = D(y, x) \geq 0$.
(2) $D(x, y) = 0$ if and only if $x = y$.
(3) For arbitrary $x, y, z \in \Omega$, $D(x, y) + D(y, z) \geq D(x, z)$.
A set Ω with a metric is called a metric space. The set of open neighborhoods
of a point $x \in \Omega$ is defined by $\{U_\epsilon(x); \epsilon > 0\}$ where

$$U_\epsilon(x) = \{y \in \Omega \ ; \ D(x, y) < \epsilon\}.$$

A metric space Ω is called separable if there exists a countable and dense
subset $\{x_i; i = 1, 2, 3, ...\}$. A set $\{x_i; i = 1, 2, 3, ...\}$ is said to be a Cauchy
sequence if, for arbitrary $\delta > 0$, there exists M such that

$$i, j > M \Longrightarrow D(x_i, x_j) < \delta.$$

If any Cauchy sequence in a metric space Ω converges in Ω, then Ω is called
a complete metric space.

Example 71. (1) Finite dimensional real Euclidean space $\mathbb{R}^d$ is a separable
and complete metric space with the metric

$$D(x, y) = |x - y| \equiv \left(\sum_{i=1}^{d} (x_i - y_i)^2 \right)^{1/2},$$

where $x = (x_i)$, $y = (y_i)$, and $|\cdot|$ is a norm of $\mathbb{R}^d$.
(2) Let K be a compact subset in $\mathbb{R}^d$. The set of all continuous function
from $\mathbb{R}^d$ to $\mathbb{R}^{d'}$

$$\Omega = \{f \ ; \ f : \mathbb{R}^d \to \mathbb{R}^{d'}\}$$

is a metric space with the metric

$$D(f, g) = \|f - g\| \equiv \max_{x \in K} |f(x) - g(x)|,$$

where $|\cdot|$ is the norm of $\mathbb{R}^{d'}$. By the compactness of K in $\mathbb{R}^d$, it is proved
that Ω is a complete and separable metric space.

Definition 31. (Probability space)   Let Ω be a separable and complete metric space. A set $\mathcal{B}$ made of subsets contained in Ω is called a sigma algebra or a completely additive family if it satisfies the following conditions.
(1) If $A_1, A_2 \in \mathcal{B}$, then $A_1 \cap A_2 \in \mathcal{B}$.
(2) If $A \in \mathcal{B}$, then $A^c \in \mathcal{B}$ (A^c is the complementary set of A).
(3) If $A_1, A_2, A_3.., \in \mathcal{B}$, then the countable union $\cup_{k=1}^{\infty} A_k \in \mathcal{B}$.
A pair of a metric space and a sigma algebra $(\Omega, \mathcal{B})$ is called a measurable space. A function P

$$P : \mathcal{B} \ni A \mapsto 0 \le P(A) \le 1$$

is called a probability measure if it satisfies
(1) $P(\Omega) = 1$.

(2) For $\{B_k\}$ which satisfies $B_k \cap B_{k'} = \varnothing \ (k \ne k')$, $P(\cup_{k=1}^{\infty} B_k) = \sum_{k=1}^{\infty} P(B_k)$.

A triple of a metric space, a sigma algebra, and a probability measure $(\Omega, \mathcal{B}, P)$ is called a probability space. If $(\Omega, \mathcal{B}, P)$ satisfies the condition that an arbitrary subset of a measure zero set is contained in $\mathcal{B}$, then it is called a complete probability space. In this book, we assume that the probability space is complete.

Remark 77. Any probability space can be made complete by extending the sigma algebra and the probability measure so that any subset contained in a measure zero set belongs to the extended algebra. The smallest sigma algebra that contains all open subsets of Ω is called the Borel field. In general the Borel field is not complete but it can be made complete by such completion procedure.

Remark 78. Let $(\mathbb{R}^N, \mathcal{B}, P)$ be a probability space, where $\mathbb{R}^N$ is the N dimensional real Euclidean space, $\mathcal{B}$ the completion of the Borel field, and P a probability distribution. If P is defined by a function $p(x) \ge 0$,

$$P(A) = \int_A p(x)dx \quad (A \in \mathcal{B}),$$

then $p(x)$ is called a probability density function.

Definition 32. (Random variable)   Let $(\Omega, \mathcal{B}, P)$ be a complete probability space and $(\Omega_1, \mathcal{B}_1)$ a measurable space. A function

$$X : \Omega \ni \omega \mapsto X(\omega) \in \Omega_1$$

is said to be measurable if $X^{-1}(B_1) \in \mathcal{B}$ for arbitrary $B_1 \in \mathcal{B}_1$. A measurable function X on a probability space is called a random variable. Sometimes X is said to be an Ω_1-valued random variable. By the definition

$$\mu(B_1) = P(X^{-1}(B_1)), \tag{10.4}$$

μ is a probability measure on $(\Omega_1, \mathcal{B}_1)$, hence $(\Omega_1, \mathcal{B}_1, \mu)$ is a probability space. The probability measure μ is called a probability distribution of the random variable X. Then X is said to be subject to μ. Note that μ is the probability distribution on the image space of a function of X. The equation (10.4) can be rewritten as

$$\int_{B_1} \mu(dx) = \int_{X^{-1}(B_1)} P(da).$$

Remark 79. (1) In probability theory, the simplified notation

$$P(f(X) > 0) \equiv P(\{\omega \in \Omega; f(X(\omega)) > 0\})$$

is often used. Then by definition, $P(f(X) > 0) = \mu(\{x \in \Omega_1; f(x) > 0\})$.
(2) In descriptions of definitions and theorems, sometimes we need only the information of the image space of a random variable X and the probability distribution P_X. In other words, there are some definitions and theorems in which the explicit statement of the probability space $(\Omega, \mathcal{B}, P)$ is not needed. In such cases, the explicit definition of the probability space is omitted, resulting in the statement such as "for Ω_1-valued random variable X which is subject to a probability distribution P_X satisfies the following equality...."

Definition 33. (Expected value) Let X be a random variable from the probability space $(\Omega, \mathcal{B}, P)$ to $(\Omega_1, \mathcal{B}_1)$ which is subject to the probability distribution P_X. If the integration

$$\mathbb{E}[X] = \int X(\omega) P(d\omega) = \int x\, P_X(dx)$$

is well defined and finite in Ω_1, $\mathbb{E}[X] \in \Omega_1$ is called the expected value, the average, or the mean of X. Let S be a subset of Ω_1. The expected value with restriction S is defined by

$$\mathbb{E}[X]_{\{S\}} = \int_{X(\omega) \in S} X(\omega) P(d\omega) = \int_{x \in S} P_X(dx).$$

Remark 80. The following are elemental remarks.

(1) Let $(\Omega_1, \mathcal{B}_1)$ and X be the same as in Definition 33 and $(\Omega_2, \mathcal{B}_2)$ be a measurable space. If $f : \Omega_1 \to \Omega_2$ is a measurable function, $f(X)$ is a random variable on $(\Omega, \mathcal{B}, P)$. The expected value of $f(X)$ is equal to

$$\mathbb{E}[f(X)] = \int f(X(\omega))P(d\omega) = \int f(x)\, P_X(dx).$$

This expected value is often denoted by $\mathbb{E}_X[f(X)]$.

(2) Two random variables which have the same probability distribution have the same expected value. Hence if X and Y have the same probability distribution, we can predict $\mathbb{E}[Y]$ based on the information of $\mathbb{E}[X]$.

Definition 34. (Convergence of random variable) Let $\{X_n\}$ and X be a sequence of random variables and a random variable on a probability space $(\Omega, \mathcal{B}, P)$, respectively.

(1) It is said that $X_n \to X$ almost surely (almost everywhere), if

$$P\Big(\{\omega \in \Omega \;;\; \lim_{n \to \infty} X_n(\omega) = X(\omega)\} \Big) = 1.$$

(2) It is said that $X_n \to X$ in the mean of order $p > 0$, if $\mathbb{E}[|X_n|^p] < \infty$ and if

$$\lim_{n \to \infty} \mathbb{E}[(X_n - X)^p] = 0.$$

(3) It is said that $X_n \to X$ in probability, if

$$\lim_{n \to \infty} P(D(X_n, X) > \epsilon) = 0$$

for an arbitrary $\epsilon > 0$, where $D(\cdot, \cdot)$ is the metric of the image space of X.

(4) It is said that $X_n \to X$ in distribution or in law, if

$$\lim_{n \to \infty} \mathbb{E}[F(X_n)] = \mathbb{E}[F(X)]$$

for an arbitrary bounded and continuous function F.

Definition 35. (1) It is said that $\{X_n\}$ is uniformly tight or bounded in probability, if

$$\lim_{M \to \infty} \sup_n P(|X_n| > M) = 0.$$

(2) It is said that $\{X_n\}$ is asymptotically uniformly integrable, if

$$\lim_{M \to \infty} \sup_n \mathbb{E}[|X_n|]_{\{|X_n| > M\}} = 0.$$

If $\{X_n\}$ is asymptotically uniformly integrable, then it is uniformly bounded.

The following mathematical relations are important and useful.

Lemma 27. *(Relation between several convergences)*
(1) If $X_n \to X$ almost surely, then $X_n \to X$ in probability.
(2) If $X_n \to X$ in the mean of order $p > 0$, then $X_n \to X$ in probability.
(3) If $X_n \to X$ in the mean of order $p > 1$ and $p > q \geq 1$, then $X_n \to X$ in the mean of order q.
(4) If $X_n \to X$ in probability, then $X_n \to X$ in distribution.
(5) If $X_n \to a$ in distribution where a is a constant, then $X_n \to a$ in probability.
(6) If $X_n \to X$ in distribution, then $\{X_n\}$ is uniformly tight.
(7) If $\{X_n\}$ is uniformly tight, then there exists a subsequence of $\{X_n\}$ which converges in distribution.

In probability theory, many theorems between random variables are derived. The following are main results which we use in this book.

Lemma 28. *(Continuous mapping theorem) Assume that f is a continuous function from a metric space to a metric space. Then by the continuous mapping theorem, the following hold.*
(1) If $X_n \to X$ almost surely, then $f(X_n) \to f(X)$ almost surely.
(2) If $X_n \to X$ in probability, then $f(X_n) \to f(X)$ in probability.
(3) If $X_n \to X$ in distribution, then $f(X_n) \to f(X)$ in distribution.
Note that these results hold even if the set of discontinuous points of f is a measure zero subset.

Lemma 29. *(Convergences of synthesized random variables)*
(1) If both $X_n \to X$ and $Y_n \to Y$ almost surely, then both $X_n + Y_n \to X + Y$ and $X_n Y_n \to XY$ almost surely.
(2) If both $X_n \to X$ and $Y_n \to Y$ in the mean of order p, then $X_n + Y_n \to X + Y$ in the mean of order p.
(3) If both $X_n \to X$ and $Y_n \to Y$ in probability, then both $X_n + Y_n \to X + Y$ and $X_n Y_n \to XY$ in probability.
(4) If both $X_n \to X$ and $Y_n \to a$ in distribution where a is a constant, then $X_n + Y_n \to X + a$ and $X_n Y_n \to aX$ in distribution. Moreover if $a \neq 0$, then $X_n/Y_n \to X/a$ in distribution.

Remark 81. Even if both $X_n \to X$ and $Y_n \to Y$ in distribution, neither $X_n + Y_n \to X + Y$ nor $X_n Y_n \to XY$ in distribution, in general.

Lemma 30. *(Convergences of expected values) In order to prove the convergence of expected values, the following theorems are employed.*

(1) If $X_n \to X$ in the mean of order p, then $\mathbb{E}[X^p] < \infty$ and $\mathbb{E}[(X_n)^p] \to \mathbb{E}[X^p]$.

(2) If $X_n \to X$ almost surely, and if there exists a random variable Y such that $|X_n| < Y$ and $\mathbb{E}[Y] < \infty$, then $\mathbb{E}[X_n] \to \mathbb{E}[X]$.

(3) Assume that $\mathbb{E}[X_n]$ and $\mathbb{E}[X]$ are finite. If $X_n \to X$ in distribution and if $\{X_n\}$ is asymptotically uniformly integrable, then $\mathbb{E}[X_n] \to \mathbb{E}[X]$.

(4) If $\sup_n \mathbb{E}[|X_n|^{1+\epsilon}] < \infty$ for some $\epsilon > 0$, then $\{X_n\}$ is asymptotically uniformly integrable.

For a short description, we adopt the following definition.

Definition 36. Let $k > 0$ be a positive real value and let X and $\{X_n\}$ be random variables. It is said that $\{X_n\}$ satisfies the asymptotic expectation condition with index k, if the convergence in distribution $X_n \to X$ holds and if there exists $\varepsilon > 0$ such that $\mathbb{E}[|X|^{k+\varepsilon}] < \infty$ and

$$\sup_n \mathbb{E}[|X_n|^{k+\varepsilon} < \infty.$$

If $\{X_n\}$ satisfies the asymptotic expectation condition with index k, then $\mathbb{E}[(X_n)^k] \to \mathbb{E}[X^k]$.

The following notations are often used in statistics.

Definition 37. Let $\{a_n; a_n > 0\}$ and $\{x_n\}$ be sequences of real values.
(1) The notation

$$x_n = o(a_n)$$

means that $x_n/a_n \to 0$.
(2) The notation

$$x_n = O(a_n)$$

means that $\sup_n |X_n/a_n| < \infty$.

Definition 38. Let $\{a_n; a_n > 0\}$ and $\{X_n\}$ be sequences of real values and random variables respectively.
(1) The notation

$$X_n = o_p(a_n)$$

means that $X_n/a_n \to 0$ in probability.
(2) The notation

$$X_n = O_p(a_n)$$

means that $\{X_n/a_n\}$ is uniformly tight.

10.4　Empirical Process

In this section we explain empirical process theory which is necessary in statistics. Let $q(x)$ be a probability density function on $\mathbb{R}^M$ and $f(x, w)$ be an $\mathbb{R}^N$-valued function of $x \in \mathbb{R}^M$ and $w \in W \subset \mathbb{R}^d$ which satisfies

$$\int f(x, w)q(x)dx = 0.$$

Let $X_1, X_2, ..., X_n$ be random variables which are independently subject to the probability density function $q(x)$. The emprical process $\xi_n(w)$ is defined by

$$\xi_n(w) = \frac{1}{\sqrt{n}} \sum_{i=1}^{n} f(X_i, w). \tag{10.5}$$

We assume that each element of the covariance matrix

$$S(w) \equiv \int f(x, w)f(x, w)^T q(x)dx$$

is finite for an arbitrary $w \in W$. Let $\xi(w)$ $(w \in W)$ be a random process whose average and covariance matrix are zero and $S(w)$. By the central limit theorem, for each w, $\xi_n(w)$ converges to $\xi(w)$ in distribution. In statistics, we need stronger results than the convergence in distribution $\xi_n(w) \to \xi(w)$ for each w. For example, convergences in distribution

$$\sup_{w \in W} \|\xi_n(w)\| \to \sup_{w \in W} \|\xi(w)\| \tag{10.6}$$

and

$$\int_W \xi_n(w)^k dw \to \int_W \xi(w)^k dw \tag{10.7}$$

for some $k > 0$ are necessary in statistical theory. The empirical process theory enables us to prove such convergences.

Let $\|g\|_\infty$ be the supremum norm of a function $g(w)$,

$$\|g\|_\infty \equiv \sup_{w \in W} \|g(w)\|,$$

and $C(W)$ be a set of all continuous functions on a compact set W,

$$C(W) = \{g(w) \text{ is continuous on } W\}.$$

Then $C(W)$ is a complete and separable metric space with the norm $\| \ \|_\infty$. Both functions

$$g \mapsto \sup_{w \in W} \|g(w)\|$$

and

$$g \mapsto \int_W g(w)^k dw$$

are continuous on $C(W)$. Hence in order to prove eq.(10.6) and eq.(10.7), it is sufficient to prove the convergence in distriution $\xi_n \to \xi$ on $C(W)$ holds.

There are several mathematically sufficient conditions which ensure $\xi_n \to \xi$ in distriution on $C(W)$ holds. The following are examples of such sufficient conditions.

(1) If $f(x, w)$ is represented by

$$f(x, w) = \sum_{j=1}^{\infty} c_j(w) f_j(x)$$

where $\sum_j |c_j(w)| < M$ for some M and $\mathbb{E}[f_j(X)] = 0$ and $\sum_j \mathbb{E}[|f_j(X)|^2] < \infty$, then $\xi_n \to \xi$ in distribution on $C(W)$ holds.

(2) Assume that $\mathbb{E}[f(X, w)] = 0$ and $\mathbb{E}[|f(X, w)|^2] < \infty$ and $w \in W \in \mathbb{R}^d$. If $f(x, w)$ is sufficiently smooth, for example,

$$\max_{|k| \leq d/2+1} \sup_{w \in W} \|\partial^k f(x, w)/\partial w^k\| < \infty$$

for an arbitrary x, where $|k| = k_1 + k_2 + \cdots + K_d$ for $k = (k_1, k_2, .., k_d)$, then $\xi_n \to \xi$ in distribution on $C(W)$ holds.

10.5 Convergence of Expected Values

Let $\xi_n(w)$ be an empirical process defined by eq.(10.5). We often need the convergence

$$\mathbb{E}[F(\xi_n)] \to \mathbb{E}_\xi[F(\xi)], \tag{10.8}$$

for a given function F on $C(W)$, where $\mathbb{E}[\ ]$ and $\mathbb{E}_\xi[\ ]$ show the expected values over $X_1, X_2, ..., X_n$ and ξ respectively. If F is a bounded and continuous function on $C(W)$, then eq.(10.8) is derived from the definition of convergence in distribution $\xi_n \to \xi$. Even if F is unbounded, if F is continuous and if there exists $\varepsilon > 0$ such that

$$\sup_n \mathbb{E}[|F(\xi_n)|^{1+\varepsilon}] < \infty, \tag{10.9}$$

then eq.(10.8) holds. If $\xi_n \to \xi$ in distribution on $C(W)$ and if there exists $\varepsilon > 0$ such that

$$\mathbb{E}[\sup_{w \in W} \|\xi_n(w)\|^{k+\varepsilon}] < \infty, \tag{10.10}$$

then it is said that $\xi_n(w)$ satisfies the *asymptotic expectation condition* with index k. If such a condition is satisfied, then

$$\mathbb{E}[\sup_{w \in W} \|\xi_n(w)\|^k] \to \mathbb{E}_\xi[\sup_{w \in W} \|\xi(w)\|^k]. \tag{10.11}$$

The following lemma shows a sufficient condition for the asymptotic expectation condition with index k.

Lemma 31. *Let* $x \in \mathbb{R}^N$, $w \subset W \subset \mathbb{R}^d$. *Assume that a function* $f(x, w)$ *is represented by*

$$f(x, w) = \sum_{j=1}^{\infty} c_j(w) f_j(x).$$

Let $k \geq 2$ *be a positive integer. The sequences* $\{c_j\}$ *and* $\{t_j\}$ *are defined by*

$$c_j = \sup_{w \in W} |c_j(w)|,$$

$$t_j = \mathbb{E}[|f_j(X)|^{k+1}].$$

If $\mathbb{E}[f_j(X)] = 0$ $(j = 1, 2, 3, ...)$, *if* $\xi_n(w) \to \xi(w)$ *in distribution on* $C(W)$, *and if*

$$\sum c_j < \infty, \quad \sum c_j t_j < \infty,$$

then the sequence

$$\sup_{w} |\xi_n(w)| = \sup_{w} \left| \frac{1}{\sqrt{n}} \sum_{i=1}^{n} f(X_i, w) \right|$$

satisfies the asymptotic expectation condition with index k.

Proof. Let $\ell = k + 1$. Since x^ℓ $(\ell > 2)$ is a convex function of x, for arbitrary $\{a_j\}$ and $\{b_j\}$,

$$\left(\frac{\sum |a_j||b_j|}{\sum |a_j|} \right)^\ell \leq \frac{\sum |a_j||b_j|^\ell}{\sum |a_j|}.$$

Hence

$$\left(\sum a_j b_j \right)^\ell \leq \left(\sum |a_j| \right)^{\ell-1} \left(\sum |a_j||b_j|^\ell \right).$$

By using this inequality,

$$
\begin{aligned}
\mathbb{E}[\sup_{w} |\xi_n(w)|^\ell] &= \mathbb{E}\sup_{w}\Big| \sum_{j} c_j(w)\Big(\frac{1}{\sqrt{n}} \sum_{i=1}^{n} f_j(X_i)\Big)\Big|^\ell \\
&\le \Big(\sum_{j} c_j\Big)^{\ell-1}\Big(\sum_{j} c_j \mathbb{E}\Big|\frac{1}{\sqrt{n}} \sum_{i=1}^{n} f_j(X_i)\Big|^\ell\Big) \\
&\le \Big(\sum_{j} c_j\Big)^{\ell-1}\ell^{\ell-1}\Big(\sum_{j} c_j \mathbb{E}[|f_j(X)|^\ell]\Big) < \infty
\end{aligned}
$$

where we used the following Lemma 32. $\qquad\square$

Lemma 32. *Let* X_1, X_2, ..., X_n *be independent random variables which are subject to the same probability distribution. Let* $k \ge 2$ *be an integer. Assume that*

$$
\mathbb{E}[|X_i|^k] < \infty, \quad \mathbb{E}[X_i] = 0.
$$

Then

$$
Y_n = \frac{1}{\sqrt{n}} \sum_{i=1}^{n} X_i
$$

satisfies that, for an arbitrary positive integer n,

$$
\mathbb{E}[(Y_n)^k] \le k^{k-1}\mathbb{E}[|X|^k].
$$

Proof. Let $i = \sqrt{-1}$ and

$$
\phi(t) = \mathbb{E}[\exp(itX/\sqrt{n})].
$$

Then

$$
\mathbb{E}[(Y_n)^r] = (\phi(t)^n)^{(r)}|_{t=0},
$$

where $(\phi(t)^n)^{(r)} = (d/dt)^r(\phi(t)^n)$. In order to prove this lemma, it is sufficient to prove that, for an arbitrary integer r ($r = 1, 2, ..., k$),

$$
\Big|(\phi(t)^n)^{(r)}|_{t=0}\Big| \le r^{r-1}A^r, \tag{10.12}
$$

where $A = \mathbb{E}[|X|^k]^{1/k}$. Let us prove eq.(10.12) by mathematical induction. For the case $r = 1$, eq.(10.12) holds. For k, assume that eq.(10.12) holds for $1 \le r \le k - 1$.

$$
\begin{aligned}
(\phi(t)^n)^{(k)} &= (n\phi(t)^{n-1}\phi(t)')^{(k-1)} \\
&= \sum_{r=0}^{k-1} \binom{k-1}{r} n(\phi(t)^{n-1})^{(r)}(\phi(t)')^{(k-1-r)}.
\end{aligned}
$$

Hence

$$\left|(\phi(t)^n)^{(k)}\right| = \sum_{r=0}^{k-1} \binom{k-1}{r} \left|(\phi(t)^{n-1})^{(r)}\right| n \left|(\phi(t))^{(k-r)}\right|$$

By using $\mathbb{E}[X] = 0$,

$$
\begin{aligned}
n(\phi(t))^{(k-r)}\big|_{t=0} &= n^{1-(k-r)/2}\mathbb{E}[X^{k-r}] \\
&\leq n^{1-(k-r)/2}A^{k-r} \leq A^{k-r},
\end{aligned}
$$

for $r = 0, 1, 2, ..., k-2$, and the assumptions of the mathematical induction

$$\left|(\phi(t)^n)^{(k)}\right| \leq A^k \sum_{r=0}^{k-1} \binom{k-1}{r}(k-1)^r = A^k k^{k-1},$$

where we used $r^{r-1} \leq (k-1)^{r-1}$ for $r = 1, 2, ..., k-1$. □

10.6   Mixture by Dirichlet Process

In this section, we introduce an inifinite mixture using Dirichlet process. Firstly, the finite mixture of normal distributions is defined as follows. In this section, if a random variable X is subject to a probability distribution P, it is denoted by

$$X \sim P.$$

Let $N(x|b)$ be a normal distribution of $x \in \mathbb{R}^M$ whose average is $b \in \mathbb{R}^M$,

$$N(x|b) = \frac{1}{(2\pi)^{M/2}} \exp\left(-\frac{\|x-b\|^2}{2}\right).$$

Firstly we represent the normal mixture defined by eq.(2.28) as a sample-generating model. For a finite positive integer K, let $a = (a_1, a_2, ..., a_K)$ be a parameter which satisfies $\sum a_j = 1$ and $a_j \geq 0$, and $b_k \in \mathbb{R}^d$. The Dirichlet distribution with index $\beta = \{\beta_k\}$

$$\mathrm{Dir}(a|\beta) = \frac{1}{z_1} \prod_{k=1}^{K} (a_k)^{\beta_k - 1}.$$

Let $\psi(b_k)$ be some prior of b_k. The normal mixture is represented as a generating model of $(X_1, X_2, ..., X_n)$,

$$
\begin{aligned}
a &\sim \varphi(a|\beta), \\
k &\sim \text{Multi}(a), \\
b_i &\sim \psi(b_k), \\
x_i &\sim N(x|b_i),
\end{aligned}
$$

where $k \sim \text{Multi}(a)$ shows that an integer k $(1 \le k \le K)$ is subject to the one-time multinomial distribution with probability $a = (a_1, a_2, ..., a_K)$.

Definition 39. (Dirichlet process) Let $(\mathbb{R}^M, \mathcal{B})$ be a measurable space of $\mathbb{R}^M$, $\alpha > 0$ be a constant, and G_0 be a measure on $(\mathbb{R}^M, \mathcal{B})$. A family of a measurable sets $\{B_j \in \mathcal{B}; j = 1, 2, ..., m\}$ which satisfies

$$
B_j \cap B_k = \varnothing \ (j \ne k), \quad \cup_j B_j = \mathbb{R}^M,
$$

is called a disjoint partition of $\mathbb{R}^M$. A probability measure-valued random variable G is said to be subject to the Dirichlet process if, for an arbitrary disjoint partition,

$$
(G(B_1), G(B_2), ..., G(B_m)) \sim \text{Dir}(a|\alpha G_0(B_1), \alpha G_0(B_2), ..., \alpha G_0(B_m)).
$$

The probability distribution of G is denoted by $DP(\alpha, G_0)$.

Then the infinite mixture model, which is formally given by $K \to \infty$ and $\beta_k = \alpha/K$, is represented by

$$
\begin{aligned}
G &\sim DP(\alpha, \psi), \\
b_i &\sim G, \\
x_i &\sim N(x|b_i).
\end{aligned}
$$

It is known that the Dirichlet process gives the discrete sum with probability one.

References

[1] H. Akaike. A new look at the statistical model identification. IEEE Trans. on Automatic Control, Vol. 19, pp. 716-723, 1974.

[2] H. Akaike. Likelihood and Bayes procedure. *Bayesian Statistics*, (Bernald J.M. eds.) University Press, Valencia, Spain, pp.143-166. 1980.

[3] S. Amari, H. Nagaoka. Methods of information geometry. AMS and Oxford University Press, Oxford, UK, 2000.

[4] H. Araki. Mathematical theory of quantum fields. International Series of Monographs on Physics, Oxford University Press, 1999.

[5] M. Aoyagi, S. Watanabe. Stochastic complexities of reduced rank regression in Bayesian estimation. Neural Networks, Vol.18, No.7, pp.924-933, 2005.

[6] M. Aoyagi. Log canonical threshold of Vandermonde matrix type singularities and generalization error of a three-layered neural network in Bayesian estimation. International Journal of Pure and Applied Mathematics. Vol. 52, pp.177-204, 2009.

[7] M. Aoyagi. A Bayesian learning coefficient of generalization error and Vandermonde matrix-type singularities. Communications in Statistics Theory and Methods. Vol. 39, pp.2667-2687, 2010.

[8] M. Aoyagi. Stochastic complexity and generalization error of a restricted Boltzmann machine in Bayesian estimation. Journal of Machine Learning Research. Vol.11, pp.1243-1272, 2010.

[9] M. F. Atiyah. Resolution of singularities and division of distributions. Communications of Pure and Applied Mathematics, Vol.13, pp.145-150. 1970.

[10] I. N. Bernstein. The analytic continuation of generalized functions with respect to a parameter. Functional Analysis and Applications, Vol.6, pp.26-40, 1972.

[11] P. Billingsley. Probability and Measure. John Wiley & Sons, Hoboken, 2012.

[12] J. E. Björk. Rings of Differential Operators. North Holland, 1970.

[13] G. Bodnár, J. Schicho, A computer program for the resolution of singularities. Progress in Mathematics, Vol. 181, pp. 231-238, Birkhäuser, 1997.

[14] D. A. Cox, J. B. Little, D. O'sea. Ideals, Varieties, and Algorithms, Third Edition. Springer, 2007.

[15] D. Dacunha-Castelle, E. Gassiat. Testing in locally conic models, and application to mixture models. Probability and Statistics, Vol. 1, pp. 285-317, 1997.

[16] M. Drton, B. Sturmfels, and S. Sullivant. Algebraic factor analysis: tetrads, pentads and beyond. Probability Theory and Related Fields, Vol. 138, pp. 463-493, 2007.

[17] M. Drton, B. Sturmfels, S. Sullivant. Lectures on Algebraic Statistics. Birkhäuser, Basel-Boston-Berlin, 2009.

[18] M. Drton. Likelihood ratio tests and singularities. Annals of Statistiscs, Vol. 37, pp. 979-1012, 2009.

[19] M. Drton, M. Plummer. A Bayesian information criterion for singular models. J. Royal Statist. Soc. B, Vol. 79, Part 2, pp. 1-38, 2017.

[20] I. Epifani, S. N. MacEachern, M. Peruggia. Case-deletion importance sampling estimators: Central limit theorems and related results. Electric Journal of Statistics, Vol. 2, pp. 774-806, 2008.

[21] M. D. Escobar, Estimating normal means with a Dirichlet process prior. Journal of the American Statistical Association Vol. 89, pp. 268-277, 1994.

[22] T. S. Ferguson. A Bayesian analysis of some nonparametric problems. Annals of Statistics, Vol. 1, pp. 209-230, 1973.

[23] T. S. Ferguson. Prior distributions on spaces of probability measures. Annals of Statistics, Vol. 2, pp .615-629, 1974.

[24] K. Fukumizu. Likelihood ratio of unidentifiable models and multilayer neural networks. Annals of Statistics, Vol. 31. No. 3, pp. 833-851, 2003.

[25] S. Geisser. The predictive sample reuse method with applications. Journal of the American Statistical Association, Vol. 70, No. 350, pp. 320-328, 1975.

[26] I. M. Gelfand and G. E. Shilov. Generalized Functions. Academic Press, San Diego, 1964.

[27] A. E. Gelfand, D. K. Dey, H. Chang. Model determination using predictive distributions with implementation via sampling-based methods. Bayesian Statistics, Vol. 4, pp. 147-167, 1992.

[28] A. E. Gelfand, D. K. Dey. Bayesian Model Choice: Asymptotics and Exact Calculations. Journal of the Royal Statistical Society Series B, Vol. 56, pp. 501-514, 1994.

[29] A. Gelman, D. B. Rubin. Inference from iterative simulation using multiple sequences. Statistical Science, Vol. 7, pp. 457-511, 1992.

[30] A. Gelman, J. B. Carlin, H. S. Stern, D. B. Dunson, A. Vehtari, D. B. Rubin. Bayesian Data Analysis, third edition. CRC Press Taylor & Francis Group, Boca Raton, 2014.

[31] A. Gelman, J. Hwang, A. Vehtari, Understanding predictive information criteria for Bayesian models. Statistics and Computing, Vol. 24, pp. 997-1016, 2014.

[32] S. J. Gershman, D. M. Blei, A tutorial on Bayesian nonparametric models. Jounal of Mathematical Psychology, Vol. 56, pp. 1-12, 2012.

[33] J. Geweke, Evaluating the accuracy of sampling-based approaches to the calculation of posterior moments. Bayesian Statistics, Vol.4, pp. 169-193. Oxford University Press, 1992.

[34] I. J. Good. Rational Decisions. Journal of the Royal Statistical Society, Series B, Vol. 14, pp. 107-114, 1952.

[35] J. A. Hartigan. A failure of likelihood asymptotics for normal mixtures. Proceedings of Barkeley Conference in Honor of J. Neyman and J. Kiefer, Vol. 2, pp. 807-810, 1985.

[36] P. Heidelberger, P. D. Welch. A spectral method for confidence interval generation and run length control in simulations. Communications of ACM, Vol. 24, pp. 233-245, 1981.

[37] H. Hironaka. Resolution of singularities of an algebraic variety over a field of characteristic zero. Annals of Mathematics, Vol. 79, pp. 109-326, 1964.

[38] M. D. Hoffman, A. Gelman. The No-U-Turn sampler: adaptively setting path lengths in Hamiltonian Monte Carlo. Journal of Machine Learning Research, Vol. 15, pp. 1593-1623, 2014.

[39] H. Ishwaran, L. F. James. Gibbs Sampling Methods for Stick-Breaking Priors. Journal of the American Statistical Association, Vol. 96, pp. 161-173, 2001.

[40] S. Janson, Gaussian Hilbert Spaces. Cambridge University Press, Cambridge, 1997.

[41] M. Kashiwara. B-functions and holonomic systems. Inventiones Mathematicae, Vol. 38, pp. 33-53, 1976.

[42] J. Kollár. Lectures on Resolution of Singularities. Princeton University Press, USA, 2007.

[43] S. Konishi and G. Kitagawa, G. Information Criteria and Statistical Modeling. Springer, 2008.

[44] F. Komaki. On asymptotic properties of predictive distributions. Biometrika, Vol. 83, No. 2, pp. 299-313, 1996.

[45] S. Lin. Asymptotic approximation of marginal likelihood integrals. arXiv:1003.5338v2, 2011.

[46] R. McElreath. Statistical Rethinking: A Bayesian Course with Examples in R and Stan. Chapman & Hall/CRC Texts in Statistical Science, 2015.

[47] S. N. MacEachern, Estimating normal means with a conjugate style Dirichlet process prior. Communications in Statistics - Simulation and Computation. Vol. 23, pp. 727-741, 1994

[48] G. McLachlan, D. Peel. Finite Mixture Models. John Wiley & Sons, New York, 2000.

[49] N. Metropolis, A. W. Rosenbluth, M. N. Rosenbluth, A. H. Teller,; E. Teller, Equations of state calculations by fast computing machines. Journal of Chemical Physics, Vol. 21, pp. 1087-1092, 1953.

[50] F. Mosteller, D. L. Wallace. Inference in an authorship problem. Journal of the American Statistical Association, Vol. 58, pp. 275-309. 1963.

[51] D. Mumford. The red book of varieties and schemes (2nd Edition). Springer-Verlag, Berlin, 1999.

[52] K. Nagata, S. Watanabe, Asymptotic behavior of exchange ratio in exchange Monte Carlo method. Neural Networks, Vol. 21, No. 7, pp. 980-988, 2008.

[53] S. Nakajima, S. Watanabe. Variational Bayes solution of linear neural networks and its generalization performance. Neural Computation. Vol. 19, No.4, pp. 1112-1153, 2007.

[54] R. M. Neal. MCMC Using Hamiltonian Dynamics. Handbook of Markov Chain Monte Carlo, Chapman and Hall/CRC. 2011.

[55] T. Oaku. Algorithms for b-functions, restrictions, and algebraic local cohomology groups of D-modules. Advances in Applied Mathematics, Vol. 19, pp. 61-105, 1997.

[56] T. Oaku. Algorithms for the b-function and D-modules associated with a polynomial. Journal of Pure Applied Algebra, Vol. 117-118, pp. 495-518, 1997.

[57] M. Peruggia. On the variability of case-deletion importance sampling weights in the Bayesian linear model. Journal of the American Statistical Association, Vol. 92, pp. 199-207. 1997.

[58] M. Plummer, N.Best, K. Cowles, K. Vines. CODA : convergence diagnosis and output anlaysis for MCMC. R News, Vol.6, No.1, pp. 7-11, 2006.

[59] M. Plummer. Penalized loss functions for Bayesian model comparison. Biostatistics, Vol.9, pp. 523-539, 2008.

[60] A. Raftery, S. Lewis. How many iterations in the Gibbs sampler? Vol. Bayesian Statistics-4, pp. 763-773. Oxford University Press, 1992.

[61] D. Ruelle. Thermodynamic Formalism. Addison Wesley, 1978.

[62] D. Rusakov, D. Geiger. Asymptotic model selection for naive Bayesian networks. Journal of Machine Learning Research. Vol. 6, pp. 1-35, 2005.

[63] M. Saito. On real log canonical thresholds.arXiv:0707.2308v1, 2007.

[64] M. Sato, T. Shintani. On zeta functions associated with prehomogeneous vector space. Annals of Mathematics, Vol. 100, pp. 131-170, 1974.

[65] G. Schwarz. Estimating the dimension of a model. Annals of Statistics, Vol. 6, No. 2, pp. 461-464. 1978.

[66] J. Sethuraman. A constructive definition of Dirichlet priors. Statistica Sinica, Vol. 4, pp. 639-650, 1994.

[67] D. Simon, A. D. Kennedy, B. J. Pendleton, D. Roweth. Hybrid Monte Carlo. Physics Letters B, Vol. 195, pp. 216-222, 1987.

[68] R. Shibata. An optimal model selection of regression variables. Biometrika, Vol. 68, pp. 45-54, 1981.

[69] K. E. Smith, L. Kahanpää, P. Kekäläinen, W. Traves. An Invitation to Algebraic Geometry. Springer, New York, 2000.

[70] D. J. Spiegelhalter, N. G. Best, B. P. Carlin, A. van der Linde. Bayesian measures of model complexity and fit. Journal of the Royal Statistical Society, Series B, Vol. 64, pp. 583-639, 2002.

[71] D. J. Spiegelhalter, N. G. Best, B. P. Carlin, A. van der Linde. The deviance information criterion: 12 years on. Journal of the Royal Statistical Society, Series B. Vol. 76, pp. 485-493, 2014.

[72] M. Stone. Cross-validatory choice and assessment of statistical predictions. Journal of the Royal Statistical Society, Series B, Vol. 36, pp. 111-147, 1974.

[73] R. H. Swendsen, J.S. Wang. Replica Monte Carlo simulation of spin glasses. Physical Review Letters, Vol. 57, pp. 2607-2609, 1986.

[74] A. Vehtari, J. Lampinen. Bayesian model assessment and comparison using cross-validation predictive densities. Neural Computation. Vol.14, pp. 2439-2468, 2002.

[75] A. Vehtari, J. Ojanen. A survey of Bayesian predictive methods for model assessment, selection and comparison. Statistics Surveys, Vol. 6, pp. 142-228, 2012.

[76] A. Vehtari, A. Gelman, J. Gabry. Practical Bayesian model evaluation using leave-one-out cross-validation and WAIC. J. Stat Comput, 2016. doi:10.1007/s11222-016-9696-4.

[77] A. W. van der Vaart, J. A. Wellner. Weak Convergence and Empirical Processes. Springer, New York, 1996.

[78] A. W. van der Vaart. Asymptotic statistics. Cambridge University Press, Cambridge, 1998.

[79] K. Watanabe, S. Watanabe. Stochastic complexities of Gaussian mixtures in variational Bayesian approximation Journal of Machine Learning Research. Vol. 7, pp. 625-643, 2006.

[80] S. Watanabe. Algebraic analysis for nonidentifiable learning machines. Neural Computation, Vol. 13, No. 4, pp. 899-933, 2001.

[81] S. Watanabe. Algebraic geometrical methods for hierarchical learning machines. Neural Networks, Vol. 14, No. 8, pp. 1049-1060, 2001.

[82] S. Watanabe. Algebraic Geometry and Statistical Learning Theory. Cambridge University Press, Cambridge, 2009.

[83] S. Watanabe. Equations of states in singular statistical estimation. Neural Networks, Vol. 23, No. 1, pp. 20-34, 2010.

[84] S. Watanabe. Asymptotic equivalence of Bayes cross validation and widely applicable information criterion in sngular learning theory. Journal of Machine Learning Research, Vol. 11, pp. 3571-3591, 2010.

[85] S. Watanabe. A widely applicable Bayesian information criterion. Journal of Machine Learning Research, Vol. 14, pp. 867-897, 2013.

[86] S. Watanabe. Bayesian cross Validation and WAIC for predictive prior design in regular asymptotic theory. arXiv:1503.07970, 2015.

[87] S. Watanabe. Theory and method of Bayes statistics. Corona publishing, Tokyo, 2012 (In Japanese).

[88] K. Yamazaki, S. Watanabe. Singularities in mixture models and upper bounds of stochastic complexity. Neural Networks. Vol. 16, No. 7, pp. 1029-1038, 2003.

[89] K. Yamazaki, S. Watanabe. Algebraic geometry and stochastic complexity of hidden Markov models. Neurocomputing. Vol. 69, pp. 62-84, 2005.

[90] K. Yamazaki, S. Watanabe. Singularities in complete bipartite graph-type Boltzmann machines and upper bounds of stochastic complexities. IEEE Transactions on Neural Networks,Vol. 16, No .2, pp. 312-324, 2005.

[91] K. Yamazaki, M. Aoyagi, S. Watanabe. Asymptotic analysis of Bayesian generalization error with Newton diagram. Neural Networks. Vol. 23, No.1, pp. 35-43, 2010.

[92] P. Zwiernik. An asymptotic behaviour of the marginal likelihood for general Markov models. Journal of Machine Learning Research. Vol. 12, pp. 3283-3310, 2011.

Index

acceptance probability, 211
AIC, 233
AICb, 233
almost surely convergence, 299
alternative hypothesis, 270
average, 6, 298
average log loss function, 68

Bayes' theorem, 7
Bayesian LASSO, 260
Bayesian model comparison, 275
BIC, 246
Borel field, 297
burn-in, 211

consistency of model selection, 250
convergence in mean, 299
convergence in probability, 299
covariance matrix, 7
critical point, 277
cross validation, 232
cross validation error, 18
cross validation loss, 18
cumulant generating function, 80
curse of dimensionality, 208
CV, 232

deep learning, 265
delicate case, 238
delta function, 5, 293
detailed balance condition, 209
DIC, 234

efficiency of model selection, 250
entropy, 68
entropy barrier, 211
equilibrium state, 5, 208
essentially unique, 69
expected value, 6

F(MCMC), 246
F(REG), 246
formal optimality, 3, 267
free energy, 21

gamma distribution, 262
generalization error, 17
generalization loss, 17

Hamiltonian, 207
Hamiltonian Monte Carlo, 211
hierarchical Bayes, 286
Hironaka theorem, 181
hypothesis test, 271

ill-posed problem, 2
importance sampling, 208
importance sampling cross validation,
    20
improper, 10
influential observation, 167
irreducible condition, 209
ISCV, 20, 232

Kullback-Leibler distance, 294

LASSO, 260

leverage sample point, 167, 243
linear regression, 48
log density ratio function, 72
loss and error, 21

MAP estimator, 197
MAP loss, 235
marginal likelihood, 10, 21
Markov chain Monte Carlo, 209
mathematical relation of priors, 254
mean, 6
measurable function, 297
measurable space, 297
metric, 296
metric space, 296
metropolis method, 209
Metropolis-Hasting Method, 210
minus log marginal likelihood, 21
ML estimator, 197
ML loss, 235
most powerful test, 272
multi-index, 136
multinomial distribution, 41
multiplicity, 147

neural networks, 53
normal crossing, 137
normal distribution, 5, 35
normal mixture, 56
normalized observable, 77
null hypothesis, 270

parallel tempering, 215
partition function, 10, 21
phase transition, 277
posterior average, 10
posterior probability density, 10
posterior variance, 10
potential barrier, 211
predictive density function, 10, 11

prior, 3, 9
probability density function, 4, 297
probability distribution, 5, 297
probability measure, 297
probability space, 297
projective space, 187

random variable, 5, 297
real log canonical threshold, 147, 184
realizable, 67
reduced rank regression, 195
redundant multi-index, 147
regular, 68
relatively finite variance, 72
replica Monte Carlo, 215
resolution of singularities, 181

sample size, 8
sampling interval, 211
separable metric space, 296
set-valued random variable, 297
sigma algebra, 297
simultaneous resolution, 188
singular fluctuation, 160
spontaneous symmetry breaking, 94
standard form, 137
statistical model, 3, 9

TIC, 233
TICb, 233
time series analysis, 171
training error, 17
training loss, 17
true distribution, 8

unbalanced case, 239
uniform distribution, 5

variance, 7

WAIC, 20, 225, 234

WAIC error, 20
WBIC, 246
widely applicable information crite-
    rion, 20

zeta function, 151

Printed in the United States
by Baker & Taylor Publisher Services